BELIEVE IT OR SNOT

Also by Dani Rabaiotti
& Nick Caruso

Does It Fart?
True or Poo?

The Definitive Field Guide to
Earth's Slimy Creatures

BELIEVE IT OR
SNOT

Written by Nick Caruso & Dani Rabaiotti
Illustrated by Ethan Kocak

Quercus

INTRODUCTION

Gunge, goo, ooze, sludge, gunk, goop, slime: no doubt at some point in your life you have come into contact with one or another of these slippery substances. But what exactly is 'slime'? The word slime can be applied to a whole host of substances that are slippery, gooey or even sticky. Essentially, slimes are viscous substances, which means they are somewhere in between a solid and a liquid. The more viscous something is, the more it holds itself in place and the less likely it is to flow. For example, water has very low viscosity, whereas custard has a much higher viscosity.

You might be most familiar with the slimy secretions made by your own body. Most of these are types of mucus – a slimy substance which does not dissolve in water, and contains specialized proteins called mucins that make the mucus thick and gel-like. One of the most obvious slimes produced by our bodies, and the substance in the title of this very book, is snot. So, what is snot? Simply put, it is the type of mucus found in your nose. It can be clear, or be yellow or greenish when there is a high concentration of white blood cells or when it is especially thick. Snot is produced by the mucous cells that line your nasal passages, and helps protect the body against disease by trapping germs. It is made up of over 90% water, and not only does it contain mucins, but also special antibodies and bacteria-fighting 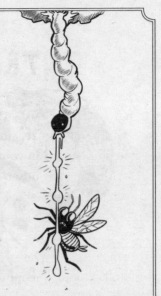 proteins that help keep us disease free. Another form of human slime you may encounter on a daily basis is saliva – also known as spit, dribble or drool. This wet substance is produced by the salivary glands in the mouth, and has an even higher water content than snot, at 99.5%. The enzymes in our saliva help kickstart the process of digestion as we chew and swallow our food.

Snot and saliva are found throughout much of the earth's biota (living organisms), but these aren't the

only substances that can make for a slimy or sticky situation. There are even some species that have their own unique slime composition, with proteins or other molecules not known to occur within any other species. Of course, these species aren't out there producing slime for the fun of it: many species' viscous secretions are not only useful but necessary to their existence – and the diversity of secretions is matched by the variety of their uses. You can learn a lot about the natural world through the slimes you find there:

plants and animals have been found to use slime for defence, respiration, movement, feeding, sending chemical signals, reproduction, hibernation and more! So, slime allows many species of animals and plants to thrive: it's the glue that holds our world together...

SLIME RATINGS

A question that many people, including us, have been seeking the answer to since the dawn of time (or at least since work started on this book) is: which animal is the slimiest? To determine just how slimy each species or group of species is, we have used our expertise in the disgusting habits of living things to devise a specialist slime-ranking system. Read on to find out who will be crowned the slimiest organism of all...

SLIME RATINGS

RATING	DESCRIPTION
0	Produces very little to no slime
✳	Makes some slime, mostly internal
✳✳	Slime is involved with this creature's daily life and can often be found on the outside of its body
✳✳✳	Touching this animal will likely leave you with a handful of slime
✳✳✳✳	Produces more slime than seems strictly necessary
✳✳✳✳✳	SO MUCH SLIME MAKE IT STOP

CONTENTS

HEDGEHOGS

Scientific name (Subfamily): Erinaceinae

RATING: ✳ ✺

Generally, if you see an animal frothing at the mouth, you should be pretty concerned, as it is often a symptom of disease. For the hedgehog, however, this behaviour is perfectly normal, and potentially even encouraged. When exposed to strong scents, hedgehogs will chew up whatever the smell in question is coming from, while producing copious amounts of saliva, turning the smelly substance into a spitty, frothy mixture. They then spread this stinky, goopy, foamy substance across their spines, a behaviour known as self-anointing. This process is repeated until much of the hedgehog is covered in smelly froth. This charming behaviour has been observed with the exposure of hedgehogs to many substances, including (but not limited to) dog poo, glue, hyacinths, cigar smoke, perfume, soap, rotting meat, fox fur, and toad skin.

There are two theories as to why hedgehogs cover themselves in slobber. One is to mask their scent and protect them from predators. The other theory comes from the fact that hedgehogs are immune to many poisons, and have been observed to self-anoint using the toxins found in toads. It is thought that hedgehogs may do this to make their spines more painful to potential predators – an extra defence keeping them safe from harm. As if you needed another reason not to touch a hedgehog.

PARCHMENT WORMS

Scientific name (Genus): *Chaetopterus*

RATING: 🔥 ✳ ⚫ 🔥

We'll cover antibacterial slime (corals, page 12), hormonal slime (snails, page 122), toxic slime (fish, page 96) and suffocating slime (hagfish, page 26) but one species has slime with a rather different characteristic – it glows in the dark. Parchment worms, in the genus *Chaetopterus*, are a type of marine polychaete worm (page 57) named after the parchment-like tubes they build and inhabit. These worms are pretty mucusy – using their slime for feeding by ejecting mucous filters from the top of their tube (similar to larvaceans, page 31) and creating a current through their parchment tube, drawing plankton and other detritus into their mucous net and consuming it.

This is far from their coolest use of parchment worm mucus though. Natural light doesn't reach the deep sea, so biolumines-cence (glowing in the dark) is important for communication, attracting prey, and more. Should a predator or intruder disturb the chimney of a deep sea parchment worm such as *Chaetopterus variopedatus*, it will emit a cloud of glowing mucus, covering the intruder and making it obvious to predators. What's amazing about the evolution of this trait is that parchment worms don't have eyes, so can't actually see the glow of their own mucus.

SLIPPERY ELMS

Scientific name (Species): *Ulmus rubra*

RATING: ✸ ☀

Plants produce a variety of compounds as a defence against herbivores, or to suppress the growth of other plants nearby which would otherwise compete for resources. Humans in turn have found an array of uses for these different compounds, such as in medicine, though not all uses are noble-minded. One plant that produces such compounds is the slippery elm, which derives its common name from its mucilaginous inner bark, and occurs naturally throughout the eastern United States and southern Canada. One of the main uses for the slippery elm's inner bark is as a demulcent, which provides soothing relief to mucous membranes, such as for a cough.

A more devious use of slippery elm was by baseball players in the early 1900s. Pitchers would chew tablets made from its inner bark to produce more slimy spit for their 'spitballs', which would cause the ball to move unpredictably. Of course, the use of these tablets and the spitball pitch are now illegal in professional baseball, although that hasn't stopped some pitchers from trying.

BIOFILMS

Scientific name (Domain): Bacteria (among others)

RATING: ✹ ✹

Biofilms are created when microorganisms, which require the aid of a microscope to be seen, stick to a surface, and then each other, through the production of slimy molecules known as Extracellular Polymeric Substances (EPS). Although biofilms are commonly associated with bacteria, other microorganisms like fungi and protists also produce these structures. For example, the dental plaque that can be found on our teeth (hopefully not much if you brush your teeth twice a day!), is a biofilm that is produced by several species of bacteria and fungi.

Biofilms can be pretty sticky too: in fact, they can attach to any surface, as long as that surface has at least some water. Within a biofilm, microorganisms can take advantage of a greater diversity of microhabitats, a higher concentration of food sources, and increased protection compared to the surrounding environment. Unfortunately, biofilms can cause problems for other species: dental plaque can cause gum disease, some biofilms provide bacteria with an increased resistance to antibiotics, and bacteria can use biofilms to colonize medical devices which can lead to harmful infectious diseases. But it isn't all bad: biofilms can be found naturally on and near roots, providing plants with necessary nutrients; and humans even use biofilms as natural filters to treat wastewater and clean up hazardous chemicals.

SIRENS

Scientific name (Family): Sirenidae

RATING: ✺ ✳ ✹

Sirens are a family of aquatic salamanders found in the south-eastern United States and northern Mexico. These unusual species lack hindlimbs, have small forelimbs relative to their body, and are the only salamanders that regularly consume plant material, although they also eat aquatic invertebrates, fish, or pretty much anything else they can catch that will fit into their mouth. Sirens also possess external gills, a trait typically found only in larval salamanders, which means they need to be in water to breathe. But like other salamanders (slimy salamander, page 88), sirens can produce a slimy coating over their body to facilitate escape from predators. We can certainly vouch for its effectiveness: sirens are tough to hold!

Sirens can often be found in temporary wetlands and have a fantastically slimy adaptation to this ephemeral lifestyle. Sirens will burrow into the mud and form a slimy, mucous cocoon around their body, similar to the lungfish (page 46), which resembles parchment when dry. Although sirens don't feed, and therefore lose body mass, while within their protective cocoon, they reduce their overall activity levels (known as aestivation) and can remain in this state for long periods of time – in laboratory conditions, one individual aestivated for over five years!

VIOLET SNAILS

Scientific name (Species): *Janthina janthina*

RATING: ✳ ✳ ✺

Like the giant African land snail (page 39) and the Antarctic limpet (page 79), the violet snail uses its mucus to stay in one place: but it doesn't attach itself to anything as solid as a rock or a house. Instead it stays at the surface of the open ocean. The violet snail accomplishes this feat by folding its mucus over air bubbles, forming a raft that is similar to bubble wrap. Of course, the ocean is constantly moving, as is the violet snail, because it drifts along with the current. If we were to use a raft we would sit on top of it, but the violet snail hangs below it, just beneath the ocean's surface.

This floating lifestyle has its challenges: if their bubble raft pops, the violet snail will sink to the bottom of the ocean and die. Individuals don't have a means to escape predators such as fish or birds, or to move around to meet a mate – instead males have to release sperm into the water in the hope that a female will come in contact with it in order to reproduce. They also can't search for food – their food has to come to them. Fortunately, food sources are relatively abundant, and these little snails will happily scrape chunks off of other organisms floating by, like Portuguese men-of-war, for meals on the go.

OPOSSUMS

Scientific name (Family): Didelphidae

RATING: ✳ ✳

Opossums are pretty weird, possessing some rather impressive anatomical oddities. First, they're marsupials, which means their young develop within the female's specialized pouch. But that's not all – females have two parallel reproductive tracts and males have bifurcated (two-pronged) genitalia to match. This unusual reproductive system led to the (wildly) incorrect belief that male opossums mated with the female's nostrils and she gave birth by sneezing into her pouch.

Although weird, none of this makes the opossum particularly slimy. However, they are periodically slimy – when they play dead. When threatened, opossums will fall to the ground with their mouth open, teeth exposed. Accompanying this award-winning acting is some slime from both ends. These marsupials will produce excessive saliva, to the point where it looks like they are foaming at the mouth, and will also produce a noxious-smelling green fluid from their anal glands. While it may be amusing to see this display (depending on your sense of humour), if you come across a wild opossum it is best to leave it alone, as their display is an involuntary fear response which can last for up to four hours, leaving them vulnerable and unnecessarily stressed.

CORALS

Scientific name (Class): Anthozoa

RATING: ✳✳

What do coral and fish (page 96) have in common? (Aside from being found in the sea; hopefully you got that one!) What many people don't know is they are both coated in mucus. Mucus is really important for protecting corals as they are sedentary and can be exposed at low tides in shallower waters. Should this happen, many corals produce extra mucus, which coats them and stops them from drying out. It also prevents other organisms growing on the coral and protects them from the sunlight, as too much UV light is damaging. Too much sunlight can cause the algae that help keep corals alive to abandon them – this is known as coral bleaching. Coral rely on the algae that live in their zooxanthellae (the individual organisms that make up a coral) to provide them with energy through photosynthesis.

Coral mucus is also important for feeding, as it encourages the growth of bacteria, which are moved with their mucus by cilia into one of the organism's many mouths, allowing the coral to digest the bacteria. Interestingly, there is also scientific evidence that the mucus creates an environment that stops bad bacteria, that would harm the coral, from growing.

MYXOZOANS

Scientific name (Class): Myxozoa

RATING: ✳ ✳

We couldn't write about slime without including myxozoans, a class of animals whose name literally means 'slime animals'. These tiny, microscopic animals have caused a lot of confusion among scientists. In the past they were considered protists – a type of organism that is not a plant, animal or a fungus – and they have even been related to slime moulds (page 92), which is how they got their slimy name. But, in more recent years, it was discovered they are actually very simple, highly specialized, jellyfish. So why did these animals evolve from the more familiar 'jelly', with its dome-like structure and long stinging tentacles?

Myxozoans are what are known as obligate parasites – that is, they have to live in other animals to survive. Most myxozoans have two hosts – worms and fish. They infect the worm and then either infect the fish when the worm is eaten, or release spores into the water, which can infect fish. This means that being tiny has major advantages, and those stinging tentacles aren't needed. Myxozoans can cause a variety of diseases in fish, including whirling disease in trout, which prevents growth and affects their ability to swim. Anglers (that is, people that fish as a hobby) can avoid spreading this disease by ensuring all their gear and boots are cleaned if they move between bodies of water.

PANGOLINS

Scientific name (Family): Manidae

RATING: ✳ ⚹

The pangolin is a pretty unusual species of mammal in that it is covered in scales made of keratin, leading them to strongly resemble a walking pine-cone. There are eight species of pangolin, family Manidae, four of which are found in Asia and four of which are found in Africa. Sadly, however, the pangolins' scales have meant that numbers have been falling rapidly, as they are used in traditional medicines. This has led to the pangolin being widely reported as the most trafficked mammal. Conservation organizations worldwide are working to stop the illegal trade in pangolin scales.

You might think that keratin scales don't sound particularly slimy, and you would be right. However, these animals do have an unusually slimy trick up their sleeve (well, mouth) – a very long sticky tongue. Pangolin saliva contains compounds that make it especially sticky, and their tongue can reach up to 70 cm long in the giant ground pangolin, *Smutsia gigantea* – that's a lot of saliva. Why the long tongue? Well, pangolins feed almost exclusively on ants and termites. Their long tongues can probe deep into ant nests and termite mounds, and their viscous saliva means the ants stick to their tongue with no hope of escape. Pangolins can consume tens of thousands of ants a day in this way!

GIRAFFES

Scientific name (Genus): *Giraffa*

RATING: ✴✴

If you think hedgehog (page 1) or guanaco (page 114) saliva is gross then you are going to be utterly horrified by the saliva of the giraffe. For one, there is a lot more of it. But it is also extra thick and sticky, and gets smeared over anything that a giraffe's tongue touches. Thanks to the giraffe's enormous neck and 21-inch-long tongue, there aren't many places the giraffe can't coat in its gloopy spit, including up its own nose – giraffes can be regularly spotted picking their nose with their blue tongue. Giraffe saliva is so thick and sticky because they feed on spiny trees and shrubs, in particular, acacia. The saliva coats the spiky twigs, protecting the giraffe's tongue and digestive tract from being stabbed by the thorns. In fact, acacia twigs can pass all the way through the giraffe's digestive system without causing any harm, although how comfortable this is for the giraffe has yet to be determined...

A team of scientists looking into what makes giraffe saliva such a good lubricant had an interesting way of collecting their giraffe spit – they headed to Edinburgh zoo with jars of apple slices. The giraffes stuck their tongues in the jars to reach the apple, coating the insides with saliva, which the scientists could then examine under a microscope. Science.

WORM-SNAILS

Scientific name (Species): *Thylacodes vandyensis*

RATING: ✳ ✳ ✳

The US Naval Ship *General Hoyt S. Vandenberg* was used as a transport ship during World War II and then as a tracking ship until 1993. So, what does a naval ship have to do with slime? Well, this ship is currently the world's second largest artificial reef, after being intentionally sunk in 2009 off the coast of Key West, Florida. This reef is home to *Thylacodes vandyensis*, a recently described species of worm-snail that, to date, has only been found on this sunken ship in the Atlantic, but is thought to be native to the Pacific Ocean.

What makes this worm-snail so spectacularly slimy is its method of hunting. Worm snails, in the Vermetidae family such as this one, 'hunt' by excreting strings from glands found at the end of two of their four tentacles. This mucous web traps food, like plankton, which can then be brought back into its mouth and the food can be extracted from the sticky trap. Unfortunately, the prevalence of *T. vandyensis* might be problematic: worm-snails produce compounds within their slime that deter many predators, allowing them to proliferate. Worm-snails bore into corals and can carry flukes, known to parasitize sea turtles (page 75).

BIVALVES

Scientific name (Class): Bivalvia

RATING: ✳ ✳ ✳

Bivalves are aquatic molluscs which have a calcium carbonate shell made up of two parts joined by a hinge. The group includes mussels, oysters, clams and scallops. Many bivalves are found at the bottom of seas, lakes or rivers; however, some species also attach themselves to rocks using specialized sticky threads.

Mucus is essential for bivalve feeding. Bivalves are filter feeders, trapping particles of food that drift into their vicinity using their gills, which have evolved into ctenidia, a specialized organ for both feeding and breathing. They trap the particles using mucus, which covers their entire body. Once particles are trapped in the mucus they are moved to the mouth using either tentacles, or specialized cilia on the gills, which waft the mucus over towards the mouth for digestion. Many bivalves also have special rod-shaped, solid mucus in their stomach, known as a crystalline style. The style pokes into a cilia lined sac next to the stomach, and these cilia rotate the style, kind of like reeling in a fishing rod, except the line is mucus and the rod is also mucus (but hard). As the style turns, the mucus is wound around it pulling in a steady stream of slime, full of tasty morsels, from the mouth into the stomach.

BLEEDING TOOTH FUNGUS

Scientific name (Species): *Hydnellum peckii*

RATING: ✺ ✺

If you live in North America or Europe, it is possible that you have seen the bleeding tooth fungus. While this mushroom does have a rather startling appearance to support its name, you might not have realized that you saw it. That's because older individuals of this species are brown and not very distinctive; but younger individuals are white and feature a dark red ooze coming out of their fruiting bodies (the fleshy portion of the mushroom), most often seen where the spore-producing structures are found. The red liquid is a by-product of excess water being drawn up through the mushroom when it grows in wet soil – the red colouration happens because the fungus naturally produces a red pigment called atromentin. It's useful stuff: due to the presence of this chemical, the red ooze may inhibit both bacterial growth and blood clotting, and can be used as a fabric dye.

Although the name 'bleeding tooth fungus' doesn't quite inspire hunger, it is edible. Unfortunately, it isn't very appetizing. Contrary to its other, more glamorous common name – 'strawberries and cream' – its taste has been described as being like a bitter pepper. Bon appétit!

SLIMEHEADS

Scientific name (Family): Trachichthyidae

RATING: ✹ ✹ ✹

Slimeheads, also known as 'roughies', are around 50 species of long-lived fish found in deep waters between 180 and 18,000 m. They have, you may not be surprised to learn, a rather slimy head, covered in canals filled with mucus. These mucous canals are part of the fishes' lateral line system, which fish use to sense water pressure and movement – if you have a pet goldfish, you can see the line of dots running down its side, also part of the lateral line system.

In the 1970s, slimeheads weren't eaten much, but thanks to advances in fishing technology, allowing vessels to reach new ocean depths, they have started to be fished. One species in particular, the orange roughy (also known as the red roughy and deep-sea perch), was particularly popular with fishing vessels. Sadly, because the orange roughy has a very slow life cycle – living up to 149 years old, and taking 30 years before it can reproduce – this meant that populations quickly plummeted, and the species was added to the Australian endangered species list. Today there has been some recovery, and some fisheries have been branded sustainable, but there's still some argument as to whether people should be eating this slimy fish.

BIRDS

Scientific name (Class): Aves

RATING: ✳

Birds, with a few exceptions (swiftlets, page 51; penguins, page 86; nighthawks, page 113) are not the slimiest of creatures. This is mostly because their feathers need to stay dry to keep them warm and work effectively for flying. They do have one pretty oily habit, however. The vast majority of birds have a gland near the base of their tail, often resembling a nipple, called an uropygial gland or preen gland, that produces an oily substance used for waterproofing and conditioning feathers. So, if you see a bird rubbing its butt, it's not just got an itchy rear – it's actually taking the oil from the preen gland to rub into its feathers.

The earliest account of this gland being discussed was by the Holy Roman Emperor, and keen falconer, Frederick II, who thought that the secretions contained a poison that owls, hawks and other birds of prey smeared on their talons to quickly kill their prey. This has since proved untrue. Still, some birds do have some other interesting uses for this oil. Hoopoes, a colourful, long-beaked bird species found across Europe and Africa, produce a particularly smelly secretion from their preen gland during the breeding season, which is often described as smelling like rotting meat. This is thought to mask the birds' scent and deter predators from their nests.

DOGS

Scientific name (Species): *Canis lupus familiaris*

RATING: ✳✳

Dogs may be man's best friend but they dribble a whole lot more than most other potential best friend candidates, (unless, that is, you are best friends with a cow, page 99). The amount of saliva produced by a dog's mouth generally depends on the size of the dog, but some dog breeds have more of a reputation for drooling than others. Due to their loose lips and large jowls, some dogs are just more prone to saliva drooling out of their mouth rather than being swallowed. Breeds such as bloodhounds, mastiffs, boxers, bulldogs and St. Bernards are notorious for leaving slimy smears on furniture.

While you have probably been dribbled on by a dog at some point in your life, what you may not know is that dog saliva has actually played a pretty important role in the science of psychology. In the 1890s, a Russian scientist named Ivan Pavlov noticed that dogs salivate when they hear someone approach with food. He wondered if he could teach them to dribble when they heard other sounds. By making a clicking sound when the dogs were fed he quickly taught the dogs to dribble at the sound of clicks, even when there was no food in sight (or smelling distance!). This is a type of learning known as classical conditioning: Pavlov's study was foundational in the field of animal behaviour as a whole.

HAGFISH

Scientific name (Family): Myxinidae

RATING: ✳︎✳︎✴︎⚫︎✳︎

If asked to think of the slimiest animals most people are probably going to think of slugs or snails (page 122), after all, their entire body is covered in mucus. But there is at least one animal that is even slimier. In fact, it produces such vast quantities of slime that when a truckload of this animal overturned in Oregon they covered an entire road, and a nearby car, with frankly excessive amounts of mucus. What weird organism is this? None other than the hagfish.

There are 76 species of hagfish, a type of jawless, eel-shaped fish that specializes in burrowing into dead or dying animals to obtain food. They are incredibly slimy – 40 mg of slime, once released from the glands along the hagfish's sides, expands over 10,000 times to produce over 1 litre of slime! Put a hagfish in a bucket and wave your hand through the water and you will pull out thick, gloopy handfuls of mucus. So why all the slime? It prevents the hagfish being eaten – if a larger fish tries to eat a hagfish it will produce enough slime to block the gills of the fish, stopping it from breathing. This elicits a rather panicked response from the would-be predator – spitting out the hagfish, which lives to slime another day.

GREAT APES

Scientific name (Family): Hominidae

RATING: ✳

When our nasal cavities get clogged with the combination of snot and debris we call bogies, we will use a tissue or handkerchief to blow our noses. Some humans (*Homo sapiens*), especially younger ones, might instead decide to use their finger to pick out the offenders, an act that is known as rhinotillexis, or maybe even eat their findings, which is known as mucophagy. It probably won't surprise you that nose-picking is pretty common among humans: one study found around 92% of people surveyed admitted to the act. Despite how common nose-picking is, this behaviour can be pretty offensive to others.

However, other great apes, like chimpanzees (*Pan troglodytes*) and western gorillas (*Gorilla gorilla*), are unfazed by onlookers while digging for and consuming nostril gold. Chimpanzees have even been observed using small sticks inside their noses to remove the unwanted material or to stimulate nerve endings found in the nasal cavity, which can cause a sneeze. But even if it was acceptable, as in other great ape species, it's best to keep your fingers out of your nose. The act of rhinotillexis can facilitate the spread of disease (viruses, page 58) or rupture small blood vessels within nasal cavities that can later become infected. Turns out your parents were right all along ...

HYENAS

Scientific name (Family): Hyaenidae

RATING: ✳✳

Hyena butter. Does that phrase spark your curiosity? Chances are if you've decided to read this book, the answer is a resounding yes! But hyena butter is not something you want to spread on a piece of toast: it is a term for the smelly, sticky, and paste-like secretion produced by the hyena's anal gland. Hyenas live in hierarchical maternal social groups, known as clans, and they use their butter to mark the clan's territory by rubbing their bums against various objects, such as branches. But since hyena clan membership is ever-changing, so is their scent. Members will show their dedication to the clan by rubbing their posteriors against already-marked territory indicators and add their own unique smell to the clan's stinky repertoire.

Through social cooperation within their clans, hyenas display a high level of intelligent behaviour, such as coordinated hunting, decision-making, and problem solving. Some studies suggest that the hyena's social intelligence may even be comparable to the intelligence of a primate. For hyenas, finding their way home doesn't require brilliance, however – they simply have to follow their noses!

LARVACEANS

Scientific name (Order): Copelata

RATING: ✳ ✳ ✳ 🖌

Larvaceans are solitary swimming tunicates – a type of filter-feeding marine animal that in its adult stage appears similar to a sponge, but when immature looks more like a see-through tadpole. Larvaceans never quite reach the 'sponge' stage, freely swimming the open oceans, known as the pelagic zone.

Giant larvaceans, such as *Bathochordaeus charon*, produce some of the largest amounts of mucus for their body size. The animals themselves aren't exactly giant – measuring around 60 mm but they produce a mucous 'house', and this can be up to a meter in diameter! These houses are made up of a coarse outer mesh and a finer inner mesh. This allows the larvacean to filter food from the water and into its mouth by creating a current through its house, which it does by beating its tail. It's kind of like living your life inside a giant fishing net made of snot, which animals stick to, then you eat them. Instead of eating their mucus like the cave glow worm (page 61), the larvacean discards its house as and when it gets clogged up, and the mucus floats down, helping take carbon to the depths of the ocean. The larvacean then simply oozes itself a new one!

STINKPOTS

Scientific name (Species): *Sternotherus odoratus*

RATING: ✷ ⚘

If you've ever held one of these turtles, you will know that both of its common names (stinkpot or common musk turtle) and its specific epithet (the second element of its scientific name – *odoratus*), are very appropriate – this is a smelly turtle. That's because the stinkpot produces a noxious and sticky substance known as musk to deter predation. Musk is produced by scent glands found on the edges of the bottom portion of the turtle's shell, which is known as the plastron.

Other reptiles, like the garter snake (*Thamnophis sirtalis*), also produce musk to make themselves less appetizing. If you've ever picked up one of these reptiles you would likely not start salivating at the thought of a delicious meal. When captured, garter snakes will twist and contort their body while producing musk from scent glands at the base of their tail (often some faeces and urine from their cloaca too), which spreads this viscous and malodorous concoction all over their body and likely onto whatever has grabbed it. But what makes this snake's musk special is its lasting power: it can often survive multiple rounds of laundering and even appears to become more pungent when rinsed with water!

HIPPOPOTAMUSES

Scientific name (Species): *Hippopotamus amphibius*

RATING: ✸ ✸

Hippos secrete a clear, viscous 'sweat' all over their body: it isn't technically sweat because these secretions are produced by subdermal glands, found beneath the skin, whereas true sweat glands, like ours, are found in the dermis. However, both of these glands act to keep the body cool by taking away body heat through evaporation. This hippo 'sweat' gradually changes colour once secreted, from clear to red, and then brown. This colour change occurs because the pigments in the 'sweat' are not stable and they quickly degrade; however, the presence of mucus can stabilize these pigments and allow the colour to remain much longer. This is beneficial to hippos because the pigmentation absorbs ultraviolet light, acting as a natural sunblock which allows hippos to stay out in the sun longer to forage for food.

Hippo 'sweat' is also acidic and acts as an antiseptic against bacterial infections, which is pretty useful because hippos are very aggressive and have the battle scars to prove it. This is all great, but we'd like to remind you: don't go trying to rub yourself on a hippo. Hippo sweat might be brilliant, but hippos are very dangerous.

SLOW LORISES

Scientific name (Genus): *Nycticebus*

RATING: ✴🦴

While venom is often associated with reptiles like rattlesnakes (Subfamily Crotalinae) or arachnids like spiders (order Araneae), there are also venomous mammals, like the five currently recognized species of slow lorises. In fact, these small fluffy animals, which somewhat resemble a bush baby if it were in the band Kiss, due to their huge eyes and dark eyepatches, are the only known venomous primates. However, unlike spiders and rattlesnakes, which have glands that secrete venom into fangs, slow lorises' venom is produced from a gland on their upper arm near the elbow, known as the brachial gland. During grooming, their oily and malodorous toxin is collected into specialized incisors known as dental combs, which allow these primates to deliver a venomous bite. Because of this, it is often debated whether or not these primates are truly venomous.

What is not debatable is that that these species can be dangerous when they feel threatened, their bite is painful and may induce anaphylactic shock in some people. While the slow loris' venom isn't quite as slimy as a lot of the other flora and fauna featured in this book, the fact that it has the potential to be fatal gives it an extra bump in our rating system. Interestingly, chemical analyses reveal that compounds contained within slow loris venom share similar genetic sequencing to cat allergens, which likely explains the variation in human reaction to their bites.

SQUID

Scientific name (Superorder): Decapodiformes

RATING: ✳ ✳ ✹ ❗

Squid really know how to make the most of their slimy secretions. Some squid use their mucus for capturing prey. The vampire squid (*Vampyroteuthis infernalis*), which is another deep-sea cephalopod, uses long, sticky filaments to capture sinking particles known as 'marine snow', which consists of decaying parts of plants and animals, faeces, and inorganic material like sand. Before ingesting the marine snow, vampire squid will also wrap up their food with their mucus.

Other squid use their secretions for defence. Squid like the deep-sea bobtail squid (genus *Heteroteuthis*) secrete a combination of ink and mucus to distract predators. This is similar to the defence mechanism seen in octopuses (page 65) although deep-sea bobtail squid also secrete luminescent bacteria, which create beautiful glowing mucusy ink clouds. Defensive secretions are also present in the wonderfully named and quite beautiful striped pyjama squid (*Sepioloidea lineolata*), which cover themselves with a layer of gel-like slime, and produce copious amounts when threatened. Although this slimy covering is similar to mucus, it doesn't contain mucins; rather, it contains numerous proteins that are only found in cephalopods. This slime also helps these squid camouflage themselves – by altering their stickiness, squid can adhere small rocks to their bodies to blend in with their surroundings!

GIANT AFRICAN LAND SNAILS

Scientific name (Species): *Achatina fulica*

RATING: ✳ ✳ ✺

As its common name suggests, this snail is native to Africa, specifically East Africa, and it is a big snail – an adult can reach nearly 20 cm in length. With that large size comes a lot of slime. Giant African land snails, like other snails (see page 122), cover their single huge foot with mucus, which is produced from their suprapedal gland, located deep in their body and above their foot. Along with communication, snail mucus reduces friction during movement which helps prevent tissue damage and allows snails to adhere or cling to structures like rocks and sticks. It also lets stationary snails form an epiphragm, which is layer of dried mucus that surrounds the opening or aperture of the shell connected to a substrate. This not only provides attachment for the snail but also helps to reduce water loss.

Unfortunately, the mucous trail of the giant African land snail can be found in a lot of places that it isn't supposed to be. The species is considered to be one of the most invasive species in the world and a major agricultural pest due to its broad range of host plants and large size – it has even been known to eat the stucco off of houses!

HORNED LIZARDS

Scientific name (Genus): *Phrynosoma*

RATING: ✳ ✳

There isn't much protecting the inside of an animal's mouth, so eating something that can bite or sting can be challenging. That's why most will only eat dangerous ant species during drought conditions, when food is scarce. However, horned lizards specialize in eating ants, which is known as myrmecophagy. This might not seem like a very nutritious meal, but horned lizards have a few adaptations which allow them to eat copious amounts of their insect prey. First, horned lizards have a natural resistance to ant venom. Second, their stomachs are disproportionately large so that they can eat a lot: horned lizards will sit next to an ant colony and devour every worker that passes by, which can lead to a very full and nearly immobile lizard.

But their most important (and slimy) adaptation is that their pharynx, the cavity connecting the mouth to the oesophagus, is lined with a thick coating of mucus-producing cells. Ants unfortunate enough to be gobbled up, are subsequently rolled up in mucus, rendering their sharp, piercing mandibles and venomous stingers ineffective. This is a very useful adaptation, as stings from some ant species like the Maricopa harvester ant (*Pogonomyrmex maricopa*), can be fatal to these lizards and other small animals.

POLYESTER BEES

Scientific name (Species): *Colletes inaequalis*

RATING: ✳ ✳

If bees are associated with any gooey, sticky substance it's probably honey, but actually, another type of bee produces something even more sticky. Polyester bees (genus *Colletes*), also known as plasterer bees and cellophane bees, are found in most of the Northern hemisphere, and there are over 160 species across Europe and North America. These small bees dig their nests in sandy, often wet, soils. The female bee will excavate a tunnel with a number of chambers leading off it and then produce a sticky substance from a specialized gland known as the Dufour's gland in her abdomen (butt). She then mixes this substance with her spit, spreads it all over the inside of her nest and, when dry, this goop forms a clear substance resembling cellophane.

Not only is the lining of the nest waterproof, allowing the bees to nest in wet soils, but it also has antibacterial properties, keeping the eggs safe from disease. The waterproofed walls maintain the perfect humidity for the developing eggs and larvae, and a number of bee species also use the chambers to store a sloppy mixture of nectar and pollen, allowing it to ferment into the bee equivalent of a soup of the day, before feeding it to their offspring.

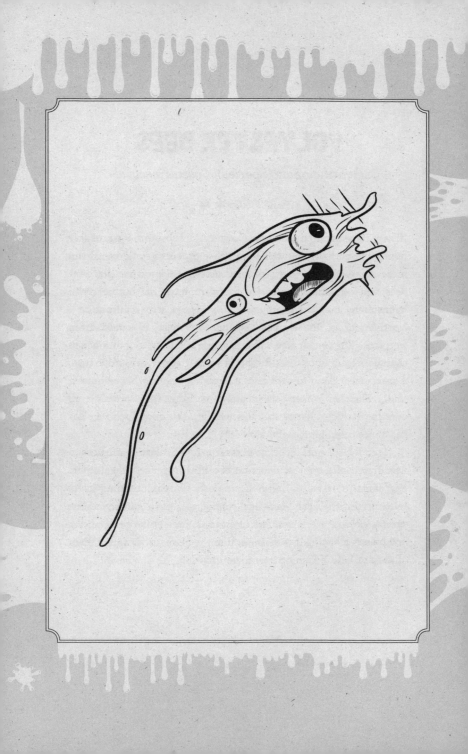

BLANKET WEED

Scientific name (Species): *Cladophora glomerata*

RATING: ✳ ✳ ✦ ▮

If you have spent a lot of time around fresh water – particularly ponds, but also rivers and streams – you may have noticed that they sometimes turn green, becoming the consistency of snot. Well, this is all thanks to a set of over 500 organisms which are colloquially known as blanket weed, or string algae, woolly algae or even horsehair. Blanket weed are algae, which might appear similar to plants – they are green and use photosynthesis – but algae do not possess a root system or flowers. Algae range in size from the microscopic to giant kelp, which reach 45 m in length. Blanket weed is a type of filamentous algae – it floats in the water without being anchored to the bottom and grows (up to 2 m a day!) in long strings.

When blanket weed takes over a pond or stream it can be bad news for the aquatic life there – it blocks out the sunlight and can starve the water of oxygen, suffocating fish and other aquatic organisms. Often this can be a sign of too many nutrients in the water, known as eutrophication, typically the result of excess fertilizer washing into the water from nearby gardens or farmland. You can prevent blanket weed from taking over by using water filters and fountains, cutting back on fertilizer in your garden and using rainwater instead of tap water to top up your pond.

SPITTLEBUGS

Scientific name (Superfamily): Cercopoidea

RATING: ✱✱✱❗

If you are out and about in the great outdoors, or even in your back garden, you may have come across white frothy blobs on some plants. Commonly referred to as cuckoo spit, other names for this sticky stuff include frog spit and snake spit. This foam doesn't actually come from a cuckoo (or frog or snake) drooling all over the place, however: it comes from froghoppers (superfamily Cercopoidea), and it isn't made up of spit at all.

Froghoppers are a small bug that get their name from their amazing ability to jump many times their own body length – up to 70 cm vertically in some species! The goopy, white mess seen on plants is actually made from foamed-up plant sap and is made by froghopper nymphs (the immature stage of many insects that have to metamorphose before they become adults and can breed), which are known as spittlebugs. This sappy, sticky home hides the nymphs from predators and parasites, stops them drying out and insulates them from both hot and cold weather. It also tastes pretty bad (we wouldn't recommend trying it), warning predators that the insects inside are unpalatable. Although these little invertebrates suck sap, most species cause little damage to the plant, and so aren't something to worry about if you have them in your garden.

LUNGFISH

Scientific name (Subclass): Dipnoi

RATING: ✹ ✹ ✹

There are six species of lungfish living on our planet today: four are found in Africa, one in South America and one in Australia. These unique animals have walked (well, swum) the planet for nearly 400 million years, with fossils dating back to the Triassic period. Unlike most fish, lungfish don't breathe primarily using gills: instead they have lungs. The clue is in the name, really. Some species will even drown if you hold them under water. So why has this evolved? Well, lungfish are the masters of living in ephemeral wetlands, where water appears seasonally.

So far, so un-slimy. But once the water levels in the habitat of the West African lungfish begin to fall during the dry season, they excavate a burrow in the mud, filling it with copious amounts of mucus, and curling up so that their head points upwards towards the entrance. Their heart rate and metabolism drop and they stop eating, in a process known as aestivation. Like sirens (page 6), the lungfish's mucus dries out once exposed to the air, forming a cocoon. Here they stay for months, until the rainy season replenishes their wetlands. In the lab, they have even been found to live up to 4 years without water, which is a pretty long time, but not so bad for a fish that has been shown to live up to 90 years!

WATER FLEAS

Scientific name (Species): *Holopedium glacialis*

RATING: ✳ ✳ ✦

As demonstrated by blanket weed (page 45), eutrophication can be disastrous for aquatic environments. However, the removal of a nutrient can also be harmful. In temperate areas of North America and Europe, the concentration of calcium in lakes has declined, especially since the 1980s, at least partly because of acid rain. This reduction in calcium has shifted the balance in water flea (order Cladocera) populations. Water fleas are small crustaceans that are important links in the food chain between plants, bacteria, and small fish. Lower calcium in these lakes means a species of water flea called *Holopedium glacialis* can better compete against water flea species within the genus *Daphnia*, which all require calcium for their carapaces, or shells.

In lieu of a carapace, *H. glacialis* is contained within a jelly capsule composed of complex sugars to protect it from predators. Unfortunately, the scales continue to tip in favour of *H. glacialis*, causing serious problems. This species is nutrient poor compared to most *Daphnia* and many species will not consume these gelatinous fleas (who would want their dinner encased in jelly?), which reduces nutrient flow through these ecosystems: sticky aggregations of *H. glacialis* can also lead to blockages in pipes and filtration systems, causing economic losses.

TARDIGRADES

Scientific name (Phylum): Tardigrada

RATING:

Tardigrades, also known as water bears or moss piglets, are a phylum of at least 900 species of microscopic animals that are found throughout the earth in both marine and freshwater habitats. While their typical habitats are rather unremarkable – they are often found in moss or lichens that are saturated with water – some species can survive the most inhospitable conditions. These slow-moving (the name tardigrade refers to their leisurely pace), eight-legged and segmented animals can survive temperatures ranging from -272°C to 150°C. Some have even survived after ten days in the vacuum of outer space – the only animal known to have accomplished this!

So how do these intrepid little creatures survive these seemingly lethal conditions? Well, unlike the lungfish (page 46) or sirens (page 6) which protect themselves within a mucous cocoon during adverse conditions, tardigrades enter a 'tun' state, which means that they retract their arms and legs, curl up, discard almost all water from their body, and reduce their metabolism to only a tiny fraction of normal functioning. While scientists are still uncovering more secrets to the tardigrades incredible ability to survive, it would seem that a dried husk-like tardigrade in its tun state is quite possibly the least slimy animal on Earth (and the wider galaxy too)!

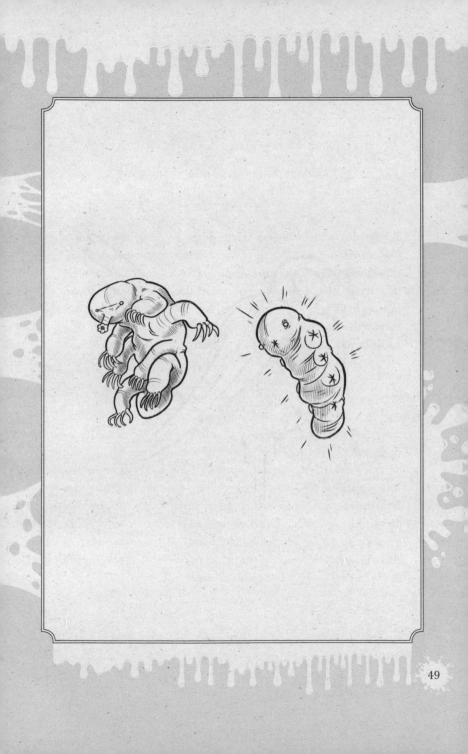

SWIFTLETS

Scientific name (Genus): *Aerodramus*

RATING: ✳ ✳

Most birds either build a nest out of twigs, feathers and other materials, or lay their eggs on the bare ground or rocks. However, for many swiftlets (genus *Aerodramus*), a type of small, fast-moving bird, building a nest proves a little tricky, as they live in caves with steep vertical walls and very little nesting material. A number of swiftlet species, including the edible-nest swiftlet, have evolved a very gooey method of dealing with the inhospitable nature of these caves – they build their nests out of saliva!

If we were to lick a rock for weeks on end all we would end up with is a very sore tongue, but for the female swiftlets this is how they construct a safe place to raise their offspring. Around the breeding season the swiftlet's salivary glands become enlarged and increase saliva production. For up to a month, the female then flies back and forth placing saliva strands onto the rock's surface, where they dry, working her way along until she has constructed a nest about the size of a teacup, resembling half a bowl stuck onto the wall, in which to lay her eggs. These nests are prized in China as an ingredient in bird's nest soup, where the nests are removed (when there aren't eggs in them!) and boiled in water to form a gelatinous soup that has been enjoyed for over 400 years.

WHALES

Scientific name (Infraorder): Cetacea

RATING: ✳

Whales aren't particularly slimy. Although they live in the water, their skin isn't covered in a thick layer of mucus, like fish (page 96). As with other mammals, however, their respiratory system does contain mucus and when whales come to the surface of the water to breathe, they rocket their snot high into the air as they exhale through their blowhole.

Marine biologists have found a way to use this snot to monitor whale species and their health. The mucus is analysed in the lab and provides DNA, hormone levels and bacteria, among other things. First, however, the scientists have to catch it. In the past this was done with a petri dish on the end of a pole – the researchers would have to drive a boat towards the whale and then cut the engine to drift close enough to get the pole under the fountain of whale snot as the whale exhaled. One group of researchers has pioneered a new method, which is less stressful for whales and less gross for the researchers – they use drones. The drone, with a petri dish attached, is flown into the whale's breath, and brings back the snot sample to researchers waiting on a boat nearby.

IRIS BORER MOTHS

Scientific name (Species): *Macronoctua onusta*

RATING: ✳✳

Gardeners are probably very familiar with the iris borer moth. As their name suggests they use irises (genus *Iris*) to complete their life cycle. Unfortunately, these moths, or more specifically the caterpillar stage of this moth, can devastate irises, which means that they are often labelled as a pest. Iris borer moths are naturally found throughout much of the eastern United States and into southern Canada. During the autumn, the adults will lay their eggs on dead iris leaves. After overwintering, the larvae hatch in the spring, coinciding with iris leaf growth – and that's when the slimy stuff starts to happen.

Iris borer caterpillars eat a tunnel through the leaves and continuously excrete waste known as frass, which can be surprisingly slimy, especially under humid conditions. This leaf damage can also cause iris leaves to release sap, a nutrient-rich fluid that can be sticky due to the presence of sugars made by plants themselves during photosynthesis. As the moth larvae mature in the summer they bore into the irises rhizome, which is the underground portion of the stem that produces new roots and shoots, and this burrowing allows bacteria to colonize and rot rhizomes giving *them* a slimy composition. The moth caterpillars then move to the soil, pupate, and adults emerge in the autumn, thus completing the iris borers' slimy life cycle!

GENETS

Scientific name (Genus): *Genetta*

RATING: ✳️ ☀️

There are currently fourteen recognized species within the genus *Genetta*, which are small omnivorous mammals that are native to Africa but have been introduced to Mediterranean Europe as well. Although they superficially resemble cats, genets are in the family Viverridae, along with civets and binturongs, and are more closely related to the mongoose. A common characteristic for members of this family is the production of a foul-smelling, thick, yellow, greasy paste from a gland near their anus, known as the perineal gland, which they use for scent marking.

What is especially notable about these secretions isn't necessarily their stickiness but rather their tendency to stick around. This might sound unappealing but you may have put these secretions on yourself: some perfumes are made from the extract of civet musk. Genet musk, however, might be in a class of its own concerning its ability to linger. One scientist, while collaring genets for a research project, allowed a genet to recover from anaesthesia in their truck; unfortunately, this individual left behind more than memories, and the putrid, musky smell remained for six months! This might not seem particularly romantic to you, but for the biologist in question it only solidified their relationship as their partner was not put off by the stench, leading to a 40-year marriage. It seems that love is not only blind, but anosmic (has no sense of smell).

VELVET WORMS

Scientific name (Class): Udeonychophora

RATING: ✸ ✷ ● ⚉

Any readers of our previous book, *True or Poo?*, will already be familiar with the soft-bodied, many legged, slime-shooting carnivore known as the velvet worm. For those who aren't, these worms immobilize their prey by shooting copious amounts of slime out of their face. The velvety texture from which the worm gets its name comes from its coverage of tiny, bristle-like papillae all over its body that make the worm water resistant and immune to its own slime, meaning it can re-ingest the slime unimpeded. Velvet worm slime is pretty incredible. For starters, slime can make up to 11% of the velvet worm's body weight. That's the equivalent of a human producing snot that weighs the same as a large cat.

Even more amazing is that the structure of their slime is not found in any other known biological substance. Whereas substances like spider silk gain their sticky property from ordered proteins, velvet worm slime is composed of unordered proteins, alongside lipids, sugars and water, giving the slime incredible elasticity and high tensile strength (it takes a lot of force to break when stretched). We know what's really on your mind though – what does it taste like? Don't worry, Henry Nottidge Moseley, a 19th-century British naturalist, has you covered. He writes in his 1874 paper that the slime has a 'slightly bitter and somewhat astringent taste'. Not for me thanks.

BRISTLE WORMS

Scientific name (Class): Polychaeta

RATING: ✹✹

Polychaete worms, or 'bristle worms', live in various watery environments and are named for the numerous hair-like structures known as chaeta. These are found at the terminal ends of muscular appendages known as parapodia, and give the worms a bristly appearance.

Bristle worms have diverse uses for their slimy secretions. In the intertidal zone you might find the mucous tracts of *Phyllodoce mucosa*, which use these slime trails to quickly find its next carrion dinner. Other species, like the aptly named sandcastle worm (*Phragmatopoma californica*) cement sand, shells and other materials together to construct a tube-home by secreting an adhesive that works even underwater and is strong enough to survive breaking waves! Polychaetes can also be found drifting in the open ocean, which can make capturing food challenging: but worms in the genus *Poeobius* come equipped with a mucous net that traps organic matter falling in the water column.

But the most extreme polychaete might just be Pompeii worms (*Alvinella pompejana*) which live near hydrothermal vents in waters as warm as 80°C. Although it isn't known for certain due to the difficulty in studying these worms in their natural habitat, scientists hypothesize that the bacteria lining them provide thermal protection, while the Pompeii worms provide their symbiotic bacteria with a nice mucusy place to live and eat.

VIRUSES

Scientific name: [None]

RATING: ✺✺

Viruses aren't technically living things (although this is perpetually up for debate) because they aren't made up of cells, and can't reproduce on their own. Viruses do, however, contain genetic material, either DNA or RNA, encased in a protein capsule, but they require their host's cells to replicate it. To increase transmission to a new host, viruses can affect their host's behaviour – like the rabies virus which causes mammals to become aggressive and more likely to bite another mammal, thus spreading the disease.

But the viruses that cause the common cold, called rhinoviruses, change their host's behaviour in a much slimier way. In humans, rhinoviruses can lead to an increased production of mucus in the nasal cavity membranes, which is known as rhinorrhea. Although mucus is present in the lining of nasal passages and airways naturally and functions as a barrier, trapping pathogens before they enter our cells, too much mucus can be a bad thing. Excess mucus helps the virus by damaging respiratory passages and the cilia lining them, which reduces the body's ability to remove pathogens and fight off infections. Moreover, viruses like the common cold have evolved to transmit through mucous droplets that are sneezed out into the environment – so always cover your nose!

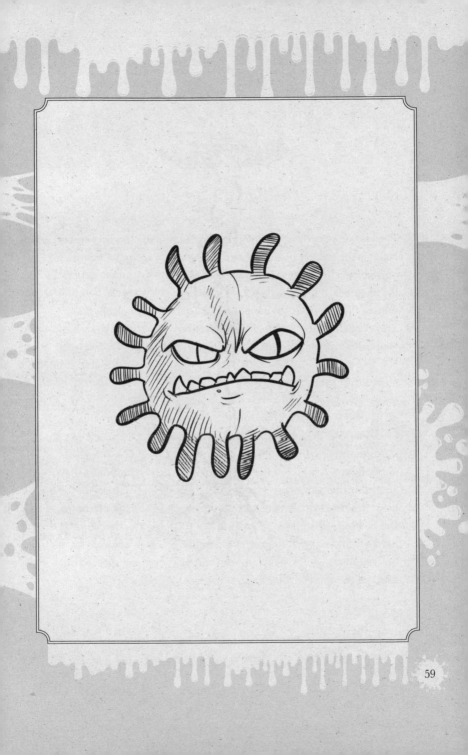

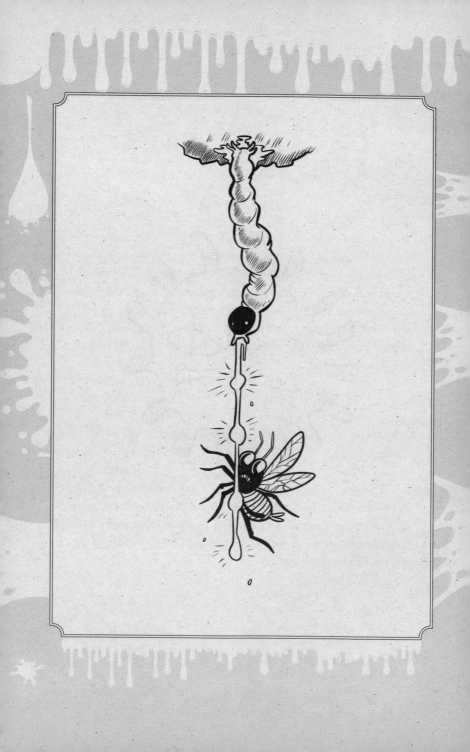

CAVE GLOW WORMS

Scientific name (Species): *Arachnocampa luminosa*

RATING: ✳ ✳ ▌

The cave glow worm, or titiwai in Maori, is a species of gnat endemic to New Zealand. Both the larvae and the adults, which only live a few days, are bioluminescent – that is, they glow in the dark. In many caves, such as the Waitomo glow worm caves on North Island, the larvae cover the cave roof, resembling stars in the night sky.

While these tiny animals look beautiful from afar, upon closer inspection you will likely find them somewhat less alluring (unless you are a flying insect). Cave glow-worm larvae hang long threads, covered in sticky urea (a waste product which we humans excrete in our urine)-filled droplets, from the cave roof to entangle unsuspecting prey. These threads are called snares, and they are up to half a metre long. Each glow worm can produce up to 70 snares – pretty impressive for an animal that is only 3 cm in length! The worms glow to attract prey – mostly midges, mosquitoes, caddisflies and moths, which see the light and think it is sunlight, and promptly fly towards the source. Once the prey touches the thread it becomes trapped in the sticky, mucus-like substance, and the worm eats the thread (waste not want not!), hauling its prey upwards in the process, before devouring it.

LANCET LIVER FLUKES

Scientific name (Species): *Dicrocoelium dendriticum*

RATING: ✹ ✷ ✸

Have you ever seen a big ball of snail mucus and thought to yourself, 'I should eat that'? Because that is exactly what species of ants in the subfamily Formicinae think. However, sometimes there is more to that ball of slime than just the usual proteins and polysaccharides that entice these ants – it may contain dozens of cercariae, the larval form of the lancet liver fluke.

Much like some viruses (page 58) the lancet liver fluke causes changes in its host to increase the probability of it being transmitted to another host – either another intermediate one (an ant again) or its primary host (typically a grazing mammal, where the parasite is able to reproduce). When the larva of this flatworm parasite is within its ant host, it changes the ant's behaviour by infecting its nervous system, causing it to remain at the top of vegetation during dusk, making it more likely to be eaten by a passing grazing animal. But when various species of terrestrial snail within the order Panpulmonata become infected by the larva, by unintentionally ingesting the eggs within piles of mammal faeces, the fluke will make its way into the snail's respiratory chamber. The presence of the fluke activates the snail's immune response which encloses the larva in a thick mucus, and, eventually, multiple larvae form a rather large slime ball, which is expelled from the snail's pneumostome (respiratory opening). And you thought your cough was bad when you had a cold!

WITCHES' BUTTER

Scientific name (Species): *Tremella mesenterica*

RATING: ✴ ✴

Tremella mesenterica, known by the common names yellow brain fungus, golden jelly fungus, yellow trembler and witches' butter, is a bright yellow jelly fungus found in deciduous woodlands on every continent other than Antarctica. Witches' butter grows on dead wood that is usually still attached to the tree. The fungus can be fairly hard to spot in dry weather, as it dries out and grows flat to the tree's surface, but after rains or damp weather the fungus produces slimy, bright yellow fruiting bodies with a jelly-like texture. These fruiting bodies are edible, and are sometimes used to give texture to soups. However, they are relatively flavourless and tricky to cook, so they aren't the most popular of mushrooms.

European folktales say that if you find witches' butter growing on your gate or door you have been cursed by a witch. The only way to lift the curse is to pierce the fungus and drain all the liquid from it. These tales are what lead to the witches' butter nickname. In reality, however, finding this fungus growing on your gate is to be fairly expected as it thrives on dead and decaying wood. It is certainly not a sign that you are cursed. We think.

OCTOPUSES

Scientific name (Order): Octopoda

RATING: ✳ ✳ ⬤ ❗

Mucus is essential to octopuses: it covers and lubricates their bodies to reduce friction and plays a key role in their defence against predation. Octopuses have an organ known as a funnel organ that secretes mucus and can be combined with their ink to produce different inky structures. For avoiding initial detection, octopuses will release ink and mucus that takes on either a rope-like appearance or that of diffuse clouds. However, when needing to escape from a predator they will create pseudomorphs, which are small mucus-rich ink clouds that resemble an octopus in shape, thus confusing the potential predator.

The southern sand octopus (*Octopus kaurna*) even uses its mucus to build a home. This octopus doesn't have the colour-changing chromatophores, or pigmented cells, that allow other octopuses to hide, so this clever cephalopod buries itself using its funnel, which propels water into the sand allowing the octopus to sink into the 'quicksand'. As it continues deeper under the surface, the southern sand octopus uses its mucus to stabilize its underground burrow and creates a chimney for ventilation. Although this sticky burrow mainly provides somewhere to hide from predators, it may also allow the octopus access to additional food sources, like worms, beneath the surface. Mucus has so many uses!

EARTHWORMS

Scientific name (Subclass): Oligochaeta

RATING: ✹ ✦ ✹ !

Earthworms are segmented worms like polychaetes (page 57), but unlike bristle worms earthworms aren't very hairy, in fact their name Oligochaeta refers to these worms having 'few hairs'. But what these worms lack in hair, they make up for in mucus, which surrounds their body and has many uses. Their mucous barrier provides the first line of defence against pathogens, allows earthworms to regulate salt and water concentrations, is used to secure their underground burrow walls, reduce friction to facilitate movement, and can even provide a protective cocoon like sirens (page 6) or lungfish (page 46) under adverse conditions. Moreover, earthworms are often associated with gardening because of their movement underground. They move soil and provide increased oxygen to plant roots, but their mucus may also provide plants with a slimy fertilizer.

Unfortunately, earthworm mucus isn't always helpful: larvae of cluster flies (genus *Pollenia*) follow the earthworm's mucous trails, burrow into their host and feed on their insides. But perhaps the slimiest earthworm is the New Zealand earthworm (*Octochaetus multiporus*). In addition to the regular earthworm mucus characteristics, this species also features a thick, bright orange-yellow bioluminescent mucus that, when disturbed, is excreted from its mouth, pores found on the underside of its body, and anus.

LEECHES

Scientific name (Class): Clitellata

RATING: ✴✴

Leeches are a group of segmented worms in the class clitellata, along with earthworms (page 66). There are over 700 species of these slimy worms globally, and while many people immediately think of parasitic leeches feeding on blood, many are predatory and feed on small invertebrates and molluscs. The medicinal leech, *Hirudo medicinalis*, was used to draw blood from patients from the 6th century until the 19th century. Leeches have a specialized anticoagulant saliva, which prevents blood from clotting so that it flows into their mouth. They are still used in other surgical procedures today, particularly in reconstructive surgery because their anticoagulant saliva can encourage blood flow to human tissue.

Leech bodies are covered in a layer of mucus, which helps them stick to the animal on which they are attempting to feed, making them pretty slimy, to which anyone who has found a leech in their sock can attest (definitely not speaking from experience). Some leeches have some especially slimy habits. The South American freshwater leech *Tyrannobdella rex* (named after *Tyrannosaurus rex* thanks to the species' exceptionally large teeth!) is sometimes found in people's noses after swimming in the Amazon river. The leech has an affinity for snot, and it is thought it evolved to live in the noses and mouths of aquatic mammals such as dolphins and otters.

EELS

Scientific name (Order): Anguilliformes

RATING: ✹ ✹ ✹

Eels don't have the best of reputations. Many people see them as gross, slimy and disgusting, and on top of that if someone is described as 'slippery as an eel' it means they are untrustworthy! Eels' sliminess, however, may be somewhat overstated. Although – as you will know if you have ever attempted to wrangle an eel – they are quite slippery, their ability to escape people's grasp has less to do with slime and more to do with their difficult-to-hold-on-to, shape.

That isn't to say that eels aren't slimy – they are covered in a layer of mucus, as are all fish (page 96). In the case of the green moray eel (*Gymnothorax funebris*), the mucus is a yellowy colour, and although the eel's skin is brown, the yellow mucus makes the eel appear green – the distinctive bright colour from which the eel gets its name. Although eels don't produce mucus in larger quantities than other fish species, there is one time that eels are particularly slimy – and that is when they are jellied. Jellied eels is a traditional English dish that involves cooking the eel in its own slime, so that the end result is a thick jelly containing chunks of eel. Eels have another rather snotty association, too – in 2018, one was found lodged up the nostril of a Hawaiian monk seal, but no-one nose why it was up there. (Sorry.)

PARROTFISH

Scientific name (Family): Scaridae

RATING: ✹ ✸ ✺

Parrotfish are a group of marine fish in the family Scaridae, found in shallow tropical and subtropical oceans around the world. They are named parrotfish after their teeth, which are fused together to form a parrot-like beak and which they use to scrape algae off hard surfaces such as rocks and coral. There are 95 species of parrotfish globally, all of which, what with being fish (page 96), are covered in a layer of mucus.

Some species, though, such as *Chlorurus sordidus*, the daisy parrotfish, take this mucous covering to the next level. Each night these fish produce a sac of mucus from special glands in their gills and burp it out, so that it encases their body like a snotty sleeping bag. It was initially thought that it served as an early warning system from predators, alerting the parrotfish should a predator get too close. However, more recent research suggests its function is actually to protect the sleeping fish from parasites. Parrotfish with a sleeping bag have been found to collect far fewer gnathiid isopods, which are small crustaceans that feed on marine fishes' blood and tissue fluids. In the day, they get eaten by cleaner wrasse, small fish which get their food from the parasites of other marine animals. At night, while the wrasse are sleeping, however, the parrotfish has to rely on its own snot to keep it safe from harm.

TUBELIP WRASSES

Scientific name (Species): *Labropsis australis*

RATING: ✳ ✳ ✦

Living things don't like being eaten, which is why many species develop evolutionary adaptations to deter predators. For example, coral (page 12) have hard and pointed calcified skeletons that are difficult enough to bite to begin with, but some species also have cells that deliver a painful and toxic sting known as nematocysts. This doesn't sound very appetizing to you or me, or in fact most fish – only about 2% of coral-associated fish, known as corallivores, will dare to dine on these cnidarians.

One such fish, however, is the tubelip wrasse; and, most appropriately, the feeding of these unique fish involves mucus produced by both the fish and the coral. That's because the tubelip wrasse secretes a thick layer of mucus across their lips which they press against the coral. The mucous barrier likely provides protection from those stinging nematocysts, but also creates a seal so that the wrasses can suck the coral's mucus off of its body. These sloppy kissers have also been observed preferentially feeding on damaged coral parts, which tend to produce more mucus compared to areas that are unharmed. Slimy yet satisfying!

CEREAL LEAF BEETLES

Scientific name (Species): *Oulema melanopus*

RATING: ✳ ✳ ✺

Humans grow a lot of food, and unfortunately, we aren't the only ones that enjoy eating it. Many species consume our food while it is still growing: these are known as agricultural pests. While some species that remain in their native range can become pests due to an artificially high density of their food source, some of the most destructive agricultural pests are those that have been introduced into new habitats. The cereal leaf beetle was originally found in Europe and Asia but has since been introduced to North America, and is considered a pest of small grains, like barley and oats, throughout its native and non-native range.

Unlike the iris borer moth (page 53), both the adult and larval forms of the cereal leaf beetle cause damage to the crops by consuming their leaves, which reduces the plants' ability to photo-synthesize and forces them to invest more energy in growth and maintenance, rather than producing grain. These beetles are similar to the iris borer moth, however, as both insects' larval form is where you'll find the slime. Larvae of the cereal leaf beetle protect themselves from predators and water loss by covering their body with a layer of mucus and faeces, giving them a shiny black appearance. Farmers are likely familiar with this disgusting covering, as walks through infested crops can yield black slimy streaks on their clothing.

SEA TURTLES

Scientific name (Superfamily): Chelonioidea

RATING: ✷✷

The stinkpot (page 32) isn't the only turtle to make it into this book: sea turtles also have some pretty impressive slime. The oviduct (the tube through which an egg is laid) of sea turtles, like the green sea turtle (*Chelonia mydas*), secretes a mucus that protects developing young in multiple ways. After being laid, a thin layer of this mucus coats the eggs and guards them from fungal infections. But perhaps most interesting is the protection this mucus provides while still in the oviduct, where it creates a low-oxygen environment. While it may seem counterintuitive to reduce oxygen for developing young, it is the result of a fascinating evolutionary adaptation.

By limiting oxygen within the oviduct, embryonic development is paused and this affords female sea turtles more time to find and select an appropriate egg-laying habitat, based on food availability and safety. This is important as once the eggs are laid, moving or turning them can damage the developing embryo – which is why you should never disturb turtle eggs. Development restarts once the eggs experience normal oxygen levels, after they are safely laid in the nest, and aren't saturated with mucus.

SEA LIONS

Scientific name (Subfamily): Otariinae

RATING: ✶ ✶ ✶

Things can get pretty gross for people who work closely with sea lions. Along with other pinnipeds, they are said to have the worst-smelling farts in the animal kingdom and the sound of their flatulence can be quite powerful as well. But their offending excretions don't stop at gas: they are reported to be the 'snottiest' animal too. Sea lions produce mucusy tears that keep their eyes lubricated and increased production of these tears washes away salt – a pretty useful adaptation to deal with salty marine conditions.

Sea lions also have mucus-lined nasal passages, but unfortunately, unlike humans, sea lions haven't quite developed manners when it comes to their snot. Their noses can 'run' due to the production of excess mucus, and these pinnipeds will blow their noses to rid blockages, albeit not with tissues. Likewise, sea lions will rid their lungs of fluids by forcefully spitting their coughed-up mucus, sometimes achieving great distances. And when their nasal passages become irritated, causing a sneeze, not only do sea lions not cover their noses but they will not think twice about sneezing all over other sea lions, or even zookeepers.

DIK DIKS

Scientific name (Genus): *Madoqua*

RATING: 🏃

If you are looking for animals that are particularly un-slimy the four species of dik dik (genus *Madoqua*) would have to be contenders. That's because these tiny African antelopes, standing at an average of 35 cm tall, are well adapted to dry environments, conserving water and getting all their hydration via the leaves and shoots they eat. A 1997 scientific paper states that dik dik have the lowest faecal water content (that's the amount of water in their poo) and most concentrated urine of any ungulate (hooved animal) tested to date, and that includes camels. Many desert species, for which the scientific term is xerocole, are specialized to conserve water. Other animals that notably produce very little water when defecating and urinating include the wombat, kangaroo rat and desert tortoise.

Sadly, despite many scientific advances in the realm of faecal analysis since 1997, no one has done a study compiling a league table of the water content in poo, so we can't conclusively say whether the dik dik takes the crown as the least slimy mammal on earth (but see tardigrades, page 49).

ANTARCTIC LIMPETS

Scientific name (Species): *Nacella concinna*

RATING: ✹ ✺

The Antarctic limpet is most likely not a household name, unless your home happens to be in the ocean along the Antarctic peninsula – then you would be very familiar with this abundant marine gastropod. Like other snails (giant African land snail, page 39; violet snail, page 9), limpets secrete copious amounts of mucus to reduce friction during movement and to cement individuals to their substrate (the surface they live on). The Antarctic limpet's mucus has another, and very important role – it protects against freezing.

Not surprisingly, it gets very cold in these marine habitats and temperatures are regularly below freezing, which is problematic because when the water inside living tissue freezes, the formation of ice crystals can damage or dehydrate the cells. But life continues to persist in these frigid habitats. Some animals produce 'antifreeze' – proteins or other substances that reduce the temperature at which their cells freeze. Similarly, the Antarctic limpet has a high concentration of salt in its tissues, but it also gets really slimy. The mucus lining their body takes longer to freeze, which is a property of the mucus' high viscosity. We wouldn't advise covering yourself in snot to keep warm though, as human mucus does not have the same properties.

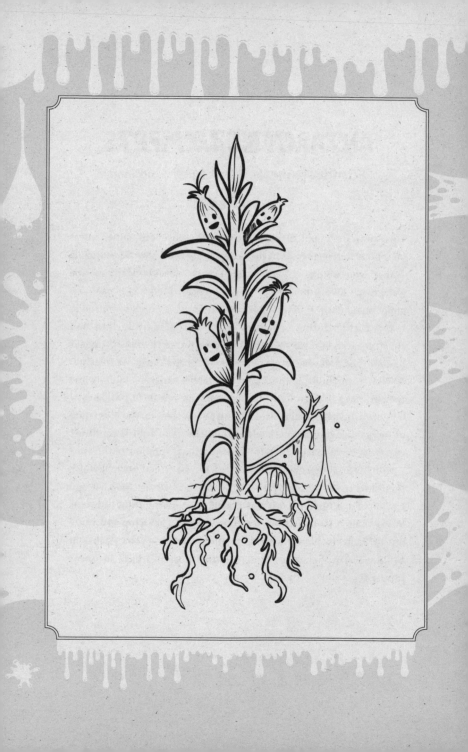

MAIZE

Scientific name (Species): *Zea mays*

RATING: ✹ ✷ ✸

If you have ever grown plants, you may have experienced root rot, whereby over-watering can cause the roots to die from lack of oxygen, or fungal attack. The telltale sign of this condition is brown and slimy roots. However, if you are in Oaxaca, Mexico and find a local variety of maize, you might notice that its roots are also slimy. These roots, however, are red or green, not brown, and are found above ground, protruding from the stalk. These slimy roots aren't a sign of disease: rather, they allow these plants to fix nitrogen from the air. Although our air is mostly composed of nitrogen (about 78%), it is unavailable to plants in this form (N_2) due to strong chemical bonds, and it typically only becomes available when converted to ammonia (NH_3) by microbes found in the soil (or unless provided fertilizer by humans).

The maize slime, which is known as mucilage, is a carbohydrate-rich viscous fluid that provides the energy source necessary to promote the growth of nitrogen-fixing microbes, letting the plant take nitrogen straight from the air. Because producing and using nitrogen-rich fertilizers leads to increased air and water pollution, scientists are currently studying these plants in hopes of transferring this slimy adaptation to other crops.

GREATER SAC-WINGED BATS

Scientific name (Species): *Saccopteryx bilineata*

RATING: ✳ ✳ ✦

The greater sac-winged bat, *Saccopteryx bilineata*, is a small bat found in Central and South America. The species gets its scientific name from the two, white zig-zag stripes on the animal's back. These bats live in groups, with one male and a number of females, known as a 'harem'. The male has a particularly gooey way of keeping the females in his harem: each day the male will mix his urine with specialized stinky secretions from gular glands under his chin, combining them in small sacs near his elbows (about halfway along the wing). He will then hover in front of the females wafting the smell from the sacs in their direction – the stinkier the better, as females find the smell incredibly alluring!

Males can spend up to an hour a day concocting the perfect perfume with which to impress the females. Should an intruder try to steal his harem, the male will aggressively fly over and flick the stinky syrup from his sacs towards his rival.

PYGMY SPERM WHALES

Scientific name (Species): *Kogia breviceps*

RATING: ✳⫰

The pygmy sperm whale (*Kogia breviceps*) is a small species of whale that only grows to about 11 ft in length. Although they are found throughout the Atlantic, Pacific and Indian Oceans, they are very elusive and rarely sighted at sea, so most scientific findings come from looking at dead whales that wash up on beaches. They are assumed to be about as slimy as most other whales (page 52) apart from one particular trait. Pygmy sperm whales, along with their close relative the dwarf sperm whale (*Kogia sima*), release a red-brown cloud into the water when startled or hunting. Previously described by scientists as 'ink' or (more … vividly) 'anal syrup', the liquid collects in a sac off the colon, and it is assumed by many to confuse predators and prey alike.

It was initially believed by scientists in the 1900s to be 'derived from the sepia of the ink-sacs of the cuttles (cuttlefish) on which the cetacean (that's a whale or dolphin) had fed'. Today it is more widely believed to be made up, at least in part, of faeces. This might be why a number of marine biologists described dead pygmy sperm whales as 'the second foulest smell they have ever had the misfortune of encountering' (after a dead leatherback turtle). No chemical analysis on the anal syrup has been reported in the scientific literature to date, so the mystery of the contents of the pygmy whale's colon sac remains unsolved. Any volunteers?

PENGUINS

Scientific name (Family): Spheniscidae

RATING: ✳

Many people who visit the zoo, or see a penguin colony in the wild, will observe the birds 'sneezing' – spraying snot with their noses accompanied by a noise akin to 'achoo'. This often gets people worried that the penguins may be ill. You can rest easy, however, because this is actually just a result of penguins' amazing adaptation for life in saltwater: penguins often take in gulps of seawater while hunting, and even those species such as the Adélie penguin (*Pygoscelis adeliae*) that get their drinking water from snow are subjected to salt water spray near the shores. Although drinking salt water can be harmful, or even deadly, to many species, this isn't the case for penguins – it doesn't ruffle their feathers! This is because penguins have specialized glands known as supraorbital glands, found just above their eye sockets, that extract excess salt from their blood. Salt is then released through nasal ducts and eventually out of their nostrils in the form of very salty snot, which they remove by shaking their heads back and forth rapidly, in what appears to be a sneeze.

But penguins aren't the only birds that tend to get a bit snotty when they take in too much saltwater. Other seabirds like shearwaters (family Procellariidae), and even some reptiles like the marine iguana (*Amblyrhynchus cristatus*), can likewise extract salt from their blood via specialized glands and excrete it, sometimes forcefully, out of their nostrils.

SHARKS

Scientific name (Superorder): Selachimorpha

RATING: ✳✳

Most sharks have fairly rough skin, which isn't particularly slimy to the touch. Of course, we do not recommend wandering around touching sharks, because they don't enjoy it very much and will likely let you know that. Most sharks have less mucus on their skin than many other types of fish, although there is still a thin layer. Scientists do, however, report that the gulper shark, *Centrophorus granulosus*, is particularly slimy relative to other sharks.

What sharks do have, however, is a lot of jelly, which has evolved as a way of detecting electrical currents in the water. The organs which do this are jelly-filled pores called ampullae of Lorenzini, which can be seen on the underside of the shark's snout – so if you see a shark with a spotty nose, it isn't a bad case of acne. When animal muscles contract it produces a tiny electrical current, and the ampullae of Lorenzini are so sensitive to electrical currents that they can detect these muscular contractions. Sharks use these electroreceptors to locate food, which is particularly useful when prey is buried under the sand.

SLIMY SALAMANDERS

Scientific name (Species): *Plethodon glutinosus* complex

RATING: ✴ ✷ ✸ ❗

Salamanders in the genus *Plethodon* are cute, abundant and diverse in the southern Appalachians: but most of all, they are slimy. The slimiest salamanders are unsurprisingly the so-called slimy salamanders, which are a complex of at least 13 similar-looking species that can be found in forested areas throughout the eastern United States. Like other salamanders in the genus, slimy salamanders produce a sticky and noxious secretion from along the top portion of their tails, and when threatened will lash their tails at whatever has grabbed them.

When handled by a human, this sticky glue-like adhesive can cause whatever the human touches next – most often leaf litter – to cling to their hands. Despite frequent and vigorous washing, these 'salamander tattoos' are often the mark of a good night catching salamanders. But when a predator is on the receiving-end of a sticky tail-whip, the viscous secretions can cause their eyes to swell shut, induce numbing and drying of the mouth, tongue swelling, and general discomfort (this was confirmed by two scientists who bravely ingested small amounts of the salamander goo). Not surprisingly, animals that wish to eat these amphibians must proceed with caution, just as the birds that have been observed pecking off their tails before dining on the rest of the salamander.

SEA HARES

Scientific name (Family): Aplysiidae

RATING: ✳✳✳❗

Unlike sea snails, which have an external shell, or nudibranchs – commonly referred to as sea slugs – which only have a shell during their larval stage, sea hares have a vestigial internal shell. Despite their name, they aren't very fast swimmers – the name 'hare' refers to their supposed resemblance, while stationary, to a sitting hare. So how do these large gastropods protect themselves against predation by other marine life? You can probably guess that, since they are featured in this book, it has something to do with sticky secretions, and you would be correct.

Like octopuses (page 65), sea hares secrete a dark ink which obscures their predator's vision but also contains chemicals that stimulate the predator's appetite so that it attacks the ink rather than the sea hare, which is known as phagomimicry. But this ink dissipates quickly leaving the sea hare vulnerable. Luckily their other secretion, which is also excreted out of its siphon, gives these gastropods a stickier solution. This whitish, opaline secretion is viscous and contains chemicals that can actually suppress the predator's appetite – likely by physically blocking chemoreceptors with stickiness, which any predator must now clean off their body. So, when a spiny lobster thinks it is going to get an easy meal, the sea hare can leave it in a cloud of ink and in serious need of grooming.

SLIME MOULDS

Scientific name (Kingdom): Protista

RATING: 🌟✴✳🍄

Despite their name, slime moulds are not, in fact, a type of mould, or even a type of fungus at all. Instead, these unique organisms, found pretty much anywhere where there is rotting wood, are a type of organism known as protists, and are sometimes dubbed 'social amoebas'. They are single-celled organisms that group together and act as one larger organism. Slime moulds' habit of moving around makes them rather tricky to keep – they send out 'pseudopods' (kind of like feelers) from lab petri dishes, and often escape their home altogether. Slime moulds can even solve mazes – by sending parts of their 'body' down different branches and retracting them when they don't find food. Pretty impressive for an organism without a brain.

Their common name certainly got one thing right, however: slime moulds are incredibly slimy. Slime moulds produce copious amounts of mucus-like slime, which coats the group of amoebas. The slime acts as 'externalized spatial memory' – that's a fancy way of saying the slime mould leaves behind a trail of slime, which tells the group where the slime mould has already been, helping it to navigate in search of food. This big, pulsating, slimy mass is so gross-looking in fact, that back in 1973, one Texas couple made headlines when a slime mould residing in their back garden was reported to be alien life. It took a mycologist (someone who studies fungi) to set the record straight.

TASMANIAN DEVILS

Scientific name (Species): *Sarcophilus harrisii*

RATING: ☀ ☀

The Tasmanian devil, *Sarcophilus harrisii,* is the largest carnivorous marsupial alive today and, as their name suggests, these black and white animals are found in Tasmania, an island off the mainland of Australia. Tasmanian devils have a ferocious reputation, in part because of their large head and formidable jaws – their bite, the strongest in relation to body weight of any animal, allows them to crush bones. In reality, however, these animals are relatively timid, and eat mostly carrion, along with wombats and small mammals.

Like all marsupials, Tasmanian devils have a mucus-lined pouch in which they raise their young. When it's time for the mother Tasmanian devil to give birth, she sticks her rear in the air and the joeys dribble down in a mucusy stream towards her backwards-facing pouch. Tasmanian devil pouch-mucus is really useful for scientists studying these animals – by examining the amount, consistency and colour of the mucus, alongside the size of the pouch, they can tell if the female is pregnant or ready to breed. This is important as these animals are endangered thanks to an infectious disease, so knowing when they are breeding helps conservationists protect these rare animals.

FISH

Scientific name (Subphylum): Vertebrata

RATING: ✸✸

If you have ever touched a fish you will probably have noticed that they are pretty slippery. It isn't just their scales, which are designed to overlap and make a fish's body as streamlined as possible, that make them feel this way, though. Over the scales, a fish's whole body is covered in a layer of mucus, which has a number of really important functions related to its survival. First, the mucus reduces drag – the friction between the water and the fish's body as it swims – which allows it to use less energy.

Fish mucus also protects them from disease. Not only does the mucus act as a barrier, preventing pathogens such as viruses and bacteria from reaching their skin, but it also contains all sorts of enzymes and proteins which kill and inhibit the growth of such bacteria. On top of that, the mucus covering the gills, with which fish breathe, helps facilitate the exchange of oxygen from the water and the excretion of carbon dioxide. In some fish, mucus plays another special role in their lives. In fish such as one species of Discus (genus *Symphysodon*), the offspring feed on the body-mucus of their parents until they are old enough to forage independently. I think we can all breathe a sigh of relief that we didn't have to consume our parents' mucus as children.

BONE-EATING WORMS

Scientific name (Genus): *Osedax*

RATING: ✳ ✳ ✳

You probably haven't come across many bone-eating worms as they are found in the benthic zone, or bottom of the ocean. Not much is known about bone-eating worms, as they have only recently been described by scientists. One of the at-least-26 species currently known to be within this genus, has the scientific name *Osedax mucofloris*, commonly translated as 'bone-eating snot-flower worm'. While the name is slimy, it would have earned a spot in this book based only on its appearance: as its name suggests, it looks like a flower made of snot! But the slime has a purpose too.

Osedax worms, also called zombie worms, feed on bones – often whale bones that have fallen to the sea floor, but it's the lipids and proteins within the bone that whets their appetite, rather than the mineral bone itself. Like other polychaete worms (page 57), zombie worms produce mucus. However, their slime is acidic, which allows them to bore into the bone and extend their 'roots' to digest and absorb the nutrients.

COWS

Scientific name (Species): *Bos taurus*

RATING: ✳✳

If you have ever spent much time around cows you will probably have noticed that they dribble a lot. Cows can produce over 180 litres of saliva *per day*! In comparison, humans produce less than 1% of that amount at 0.75–1.5 litres of saliva per day. So, what's with all the drool? Well, it's because cows, which are ruminants, have four stomach components, the first of which is the rumen. This evolved to help them break down cellulose, which is found in high quantities in grass, and is difficult for most animals to digest.

When a cow eats a mouthful of grass, the saliva and initial chewing start off the digestion process, before the food is swallowed. The food then passes into the rumen and reticulum – the first two stomach chambers, where it begins to be digested. But then, unlike what we do when we eat (I hope!), the food is regurgitated back into the mouth and chewed again; this substance is known as cud. The copious amounts of saliva help neutralize the acid from the rumen, protecting the cow's tissues. Cows can spend up to 40% of their time chewing cud, which is known as ruminating, so they need plenty of slimy spit!

STICKLEBACKS

Scientific name (Family): Gasterosteidae

RATING: ✹ ✹ ⚔

Throughout this book we have introduced animals that produce slimy or sticky secretions from various structures, but perhaps the most unusual is the stickleback's kidneys. Sticklebacks are a family of small marine and freshwater fish that have spines within their dorsal fins. After maturing, male sticklebacks increase the production of androgens, a group of hormones that regulate male reproductive characteristics, during the breeding season, and this higher concentration of hormones leads to enlarged cells within their kidneys. These cells begin producing the wonderfully named spiggin, a silk-like adhesive protein that is then stored within their urinary bladder. Male sticklebacks use this sticky protein, along with scraps of algae and other vegetation, to construct nests in sandy pits. After the female lays her eggs, the male takes on the parental care by guarding the eggs, oxygenating the water by fanning his tail, and removing any dead or diseased eggs.

While reproduction is obviously important, the male's kidneys lose the ability to filter liquid waste. To provide relief, male sticklebacks are still able to excrete liquid waste through their anal opening, but rather than their kidneys, they use specialized channels within their intestinal wall! An impressive level of commitment to parenting.

GIANT PANDAS

Scientific name (Species): *Ailuropoda melanoleuca*

RATING: ✹

The giant panda long proved a mystery to zoologists outside China. For many years, due to its weird colouration, strange diet and confusing anatomy, some scientists argued that it was not a bear, but in fact more closely related to raccoons. In fact, even up until the 1980s people were still researching whether or not the giant panda was a bear. The conclusion they came to? Yes, it is.

Pandas probably aren't the first creatures that come to mind when thinking about slime and snot – although they do produce a slimy, waxy substance from their anal glands, which as a solitary animal comes in very handy for marking your territory and letting nearby members of the opposite sex know you might be interested in reproducing. One panda in particular, however, shot to fame for its snotty habits. In 2006, a video of a baby panda at Wolong Panda Breeding Centre in China went viral. In the video, the mother panda, who is eating bamboo in the corner (as pandas have a habit of doing), jumps in fright because of her baby's very loud sneeze. To date the video has been viewed over 263 million times – a huge audience for a very tiny amount of snot!

CUSHION STARS

Scientific name (Species): *Pteraster tesselatus*

RATING: ✳ ✱ ✸ 🌢🌢

The cushion star, *Pteraster tesselatus*, also (very appropriately) known as the slime star, is a species of starfish found in the rocky areas along the coasts of the North Pacific. It can be orange, brown, grey or yellow, and has been described in publications as having five 'stumpy' arms and an 'inflated' or 'fat' appearance. There is a reason for the starfish's apparent case of bloat, however.

Their dorsal (top) surface is covered in a thick fleshy membrane which assists in the production of copious amounts of slime. The cushion star has some formidable predators, such as the morning sun star, *Solaster dawsoni*, which can easily outrun the cushion star (5 legs are no match for the morning sun star's 11–13!). When the morning sun star comes into contact with the cushion star, the cushion star envelopes itself in a thick layer of mucus, up to 7 cm deep, which prevents the tube feet of its attacker adhering to the cushion star's body. This means the morning sun star cannot climb on top of it, which, as starfishes' mouths are located under their bodies, means the morning sun star cannot feed. When studied in the lab, the slime was so good at repelling morning sun stars that it had a 100% success rate – no cushion stars were eaten after using their slimy defence.

BANANA SLUGS

Scientific name (Genus): *Ariolimax*

RATING: ✺ ✶ ✹ !

As anyone who has accidentally stepped on them barefoot can attest, slugs are very slimy. Large slugs in particular, such as the banana slug, are even slimier. There are three species of banana slug, all of which are pretty big (up to 7 inches in length!), and they can be found in the forests of the North Pacific Coast of the United States. They are so ubiquitous in these forests that they are even the mascot of the University of California, Santa Cruz.

Slime is key to slug life. The mucus they produce coats their bodies, protecting them from drying out, and it allows them to move about both on the forest floor and when climbing in search of food. Slug mucus has special properties that both adhere the slug to the surface it is moving over, and lubricate its foot (that's most of its body!) as it moves. The banana slug's uses for slime don't stop there though: their slime also contains pheromones, helping them to find other slugs with which to mate, and toxins which causes them to taste bitter to predators. This makes the annual banana-slug recipe contest, which is part of the Russian River Banana Slug Festival, even more bizarre. I think we'll skip it, thanks. Now pass the jellied eels (page 69)!

AFRICAN WILD DOGS

Scientific name (Species): *Lycaon pictus*

RATING: ✳ ❋

The African wild dog (*Lycaon pictus*) goes by many names, including, but far from limited to: painted dog, painted wolf, African hunting dog and cape hunting dog. Despite this, they are neither dogs nor wolves, although they are part of the canid family. They live in packs, with the alpha male and female having pups, and the rest of the pack helping to raise them. African wild dogs feed mostly on mammals, including impala, dik dik, wildebeest, and even baboons. Before hunting, the pack groups together in what is known as a rally – where the dogs greet each other and reinforce social bonds. Not all rallies end in hunts, however, and researchers had long wondered how exactly the dogs decide to leave.

It turns out, wild dogs have a particularly snotty method of deciding when to go hunting. Researchers noticed that pack members would always sneeze repeatedly before leaving, and that the number of sneezes needed depended on who was doing the sneezing. If it was a dominant male or female only three sneezes were needed. However, other pack members had to sneeze ten or more times to convince the pack. Whether a wild dog with a cold (viruses, page 58) adversely affects wild-dog decision making, however, has yet to be determined.

PITCHER PLANTS

Scientific name (Genus): *Nepenthes*

RATING: ✳ ❋

Nepenthes is a genus of pitcher plants found mostly in the Malaysian peninsula, with some species found as far afield as Australia, India and even Madagascar. These plants consist of shallow roots, with a number of long stems, each ending in a trap, or 'pitcher', which is filled with a syrupy liquid. Although they are often referred to as monkey cups, because primates have been observed drinking from them, the pitchers did not evolve as a form of simian pint glass. Instead, pitchers evolved to trap small animals. These plants have brightly coloured edges that attract insects and create a slippery surface, causing them to fall into the syrupy trap below. The fluid inside the pitcher contains viscoelastic polymers, which are basically special molecules that make the liquid thick, sticky and stretchy, allowing the pitcher to efficiently trap winged insects such as moths, flies and wasps.

Enzymes in the liquid both dissolve the prey and prevent rotting, allowing the plant to obtain nutrients, mostly nitrogen and phosphorus, which they can't get from the nutrient-poor soils in which they grow. It's not just insects that these plants trap in their goo, though: animals as large as lizards, rodents and even small birds have been found trapped in the liquid. And, as you may well know if you have read *True or Poo*, some pitcher plants are even specially adapted to feed off tree shrew poo.

GHOSTS

Scientific name (?): [CLASSIFIED]

RATING: ✳ ✳

In popular culture ghosts are portrayed as leaving behind a trail of slimy ectoplasm. However, it is unclear what exactly this goo is made of. The scientific definition for ectoplasm is the outer portion of the cytoplasm, that is, the material within a cell. But in terms of ghosts, ectoplasm typically refers to the physical manifestation of the contact between a spirit and a medium (a person who claims to have an ability to communicate with these apparitions). People who claim to have seen ghosts often describe ectoplasm as having a thread- or paper-like appearance. Upon closer inspection of these spectral signatures people have found nothing more than cheese-cloth or magazine cut-outs, which is likely what gives their papery appearance.

As for the slime, well ... mediums can be quite good at deception, and may hide evidence of 'ghosts', or artefacts of the dead with which they are trying to communicate, within their own body, only to regurgitate them during the ritual. This means they are covered in digestive juices and saliva. So, if there's ectoplasm in your neighbourhood, who you gonna call? Well it should probably be someone trained in hazardous waste clean-up.

FROGS

Scientific name (Order): Anura

RATING: ✹ ✷ ✸

You might be familiar with the mucous covering of some frogs, like the bullfrog (*Rana catesbeiana*), that make them slimy and often difficult to hold. This mucous layer functions to help prevent water loss due to their porous skin and even helps prevent bacterial or fungal infections. However, some species make use of additional sticky secretions. When provoked, the crucifix toad (*Notaden bennettii*), which is native to Western Australia, secretes a yellow adhesive from its back to deter predators. But this adhesive is also a useful tool to trap prey; insects trying to bite these toads will get caught in their glue trap, which the toads can later consume by shedding their skin.

Native to southern Africa, the common rain frog (*Breviceps adspersus*) also produces an adhesive, but it is used for reproduction. Typically, during frog reproduction, the male clasps the female from her back, while together they release their sperm and eggs. This isn't possible for the common rain frog – this species is too round to grab onto the females. Instead the male produces a sticky glue that allows male and female to remain attached despite their rotund shape. Their glue is so strong, scientists have reported pieces of the female's skin removed when a pair was forcefully separated!

COMMON NIGHTHAWKS

Scientific name (Species): *Chordeiles minor*

RATING: ✳ ✳

The common nighthawk, *Chordeiles minor*, is a well-camouflaged bird measuring between eight and ten inches, found across large parts of North and South America. They feed on large flying insects, such as moths, with the help of their large mouths, which are full of bristles. The birds will fly along with their mouths wide open scooping up insects in the air.

Nighthawks are ground-nesting, laying their eggs on the bare ground, using their amazing camouflage to disguise the nest from predators while the female incubates the eggs. Once the eggs hatch, both the male and female forage for flying insects, which they bring back alive to the nest to feed the offspring. When transporting the insects, the parents will carry them in their mouths. No-one wants a mouth of flappy bugs, so nighthawks coat their offspring's food in thick goo, forming the insects into what is known as a bolus. This slimy ball of writhing insects may look absolutely horrifying to us, but it looks like an amazing gourmet dinner to baby nighthawks. Scientists have been sampling these boluses (with scientific instruments, not by tasting, although you never know ... see slimy salamanders, page 88) to test what exactly this goo is made of: it is now thought to be a type of extra sticky saliva.

GUANACOS

Scientific name (Species): *Lama guanicoe*

RATING: ✳

You may not be familiar with guanacos but you've probably heard of llamas (*Lama glama*), which were domesticated from guanacos over 6,000 years ago. Guanacos are found in South America in a variety of habitats including plains, mountainous regions up to 5,000 m in elevation, and even the desert. Like the closely related llama, guanacos are herbivores, have thick woolly coats, and are generally calm and well-tempered.

These mammals have two methods of protecting themselves from predators, the first being a loud call that sounds like laughter, which alerts the rest of the herd to danger so they can run to safety. The second defence is for when guanacos feel threatened: they spit, which can be slimy, accurate, and reach a distance of up to six meters! The reason guanaco spit can be a bit slimy, and typically even more gross than other animal spit, is that it is mixed with their digestive juices and undigested food from their stomach. So as always with any wild animal it is best to keep a respectful distance, even if it appears calm, or else you might end up with guanaco gob in your face.

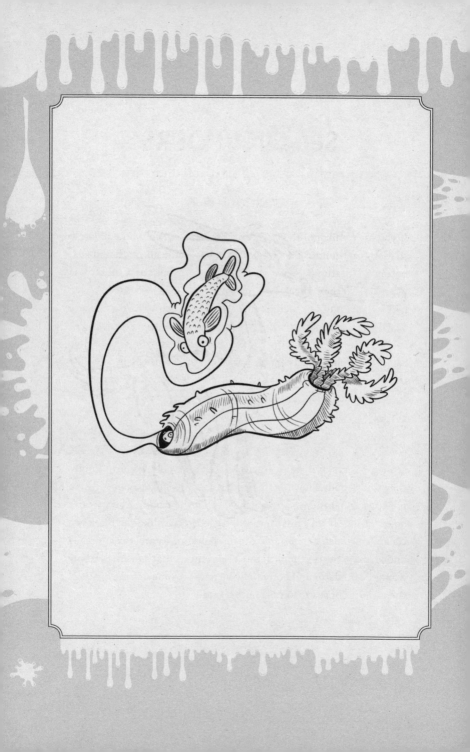

SEA CUCUMBERS

Scientific name (Class): Holothuroidea

RATING: ✳ ✳ ● ▮

If you are familiar with *Does it Fart?* or *True or Poo?* you will know that we are quite fond of sea cucumbers, and all of their fascinating adaptations. While their ability to expel their innards is noteworthy, they have slimy qualities too. Sea cucumbers have many sticky mucus-covered oral tentacles, which vary in structure based on their function: some species, like the yellow sea cucumber (*Colochirus robustus*), filter out food particles from the water column and have feathery tentacles; whereas other species, like the mucus-covered (and magnificently named) snot sea cucumber (*Leptosynapta dolabrifera*) sweep through the sandy substrate for their food and have oral tentacles that are more 'mop-like'.

But when sea cucumbers are on the defence is when things really get sticky. Preferring to avoid confrontation when possible, sea cucumbers' unique collagenous tissues enable them to liquify their entire body, allowing them to escape into small crevices where predators can't follow, and quickly stiffen their body so they can't be extracted. If they are captured, some species, like the black sea cucumber (*Holothuria forskali*) can discharge sticky and toxic threads known as Cuvierian tubules from their anuses that can ensnare any would-be predator and potentially be fatal to some marine life. Will the wonders of sea cucumbers never end?

BOX FISH

Scientific name (Species): *Ostracion cubicus*

RATING: ✳ ✳ ✦

The yellow box fish, *Ostracion cubicus*, is named after its cube-shaped body and the fact that juveniles of the species are bright yellow in colour. These charismatic little fish are found on reefs across the Pacific and Indian oceans. While having a rigid, box-shaped body and stubby little fins may not seem ideal for getting around in the ocean, these fish are surprisingly agile, and they can quickly manoeuvre around their environment despite their bizarre bodyshape. These fish are so acrobatic, in fact, that Mercedes-Benz designed a car based on their body shape.

Like all fish (page 96) the box fish's body is covered in a fine film of mucus. They have another slimy trick up their sleeve though – as with the parchment worm (page 2), these small, chunky fish use their mucus for defence. The box fish's yellow colour, combined with its black spots, are a form of aposematic colouration – the bright colours warn predators that the fish is toxic. These fish secrete a neurotoxin-filled mucus into the water when stressed, which kills other fish in the vicinity. This makes these fish rather tricky to keep in captivity, as if they become stressed they will quickly kill off their tank-mates.

UPSIDE-DOWN JELLYFISH

Scientific name (Genus): *Cassiopea*

RATING: ☀ ☀ ▮

As its name suggests, the upside-down jellyfish isn't like most other jellyfish, which have their 'bell' at the top and tentacles underneath. The currently recognized nine species in this genus all sit at the bottom of shallow ocean waters throughout the tropics and subtropics, and their oral arm-appendages, rather than tentacles, extend towards the surface, rather than the sea floor. Oral arms may look like tentacles but they differ in being highly branched, containing smaller secondary mouths along their lengths, and originating from around the jelly's mouth, near the centre of the bell, rather than around the bell's edges. Although these jellyfish contain symbiotic photosynthesizing zooxanthellae, like coral (page 12), that provide them nutrition, they also capture prey and have stinging nematocysts that both protect the jellyfish and immobilize their prey.

Unlike other jellyfish, however, the upside-down jellies will release nematocyst-filled mucus into the surrounding water, so you don't have to touch them to feel their sting! Many snorkelers have encountered this 'stinging water' and they aren't the only ones: carrier crabs (*Dorippe frascone*) have been observed carrying these jellies on their back, which likely protects and camouflages them.

SNAILS

Scientific name (Class): Gastropoda

RATING: ✳ ✳ ✳

Snails, like the giant African land snail (page 39) and their naked relatives, slugs (page 105), are known collectively as gastropods, a type of mollusc. Other than their shells, snails' most distinctive feature is probably their slime, which covers their entire body, and among other uses plays an important part in snail communication. Trails of snail mucus allow them to find their way back 'home' – that is, to safe, damp spots where groups of snails can spend time when they are not feeding.

Snails can also use their mucous trails to find other snails. This can be a good thing, such as when snails use it to find other individuals that are ready to reproduce. It has its downsides, though – although most snails eat plants, some snails, such as the rosy wolfsnail, *Euglandina rosea*, eat other snails. They will use their specialized mouthparts to follow a slimy trail, before finding and consuming their snail prey.

It's not just other snails that make use of this slime, however. Since ancient Greece, humans have been using snail slime as a facial treatment, and in recent years its use in anti-ageing creams has become increasingly common. It's not entirely clear how effective it is, however, as to date there have been no large-scale scientific studies on the topic.

AND THE WINNER IS ...

THE SLIME CHAMPION!

THE HAGFISH!

FINAL STANDINGS

Entry	Rating	Page
Hagfish		26
Slime moulds		92
Cushion stars		102
Parchment worms		2
Blanket weed		43
Larvaceans		30
Velvet worms		56
Sea cucumbers		117
Squid		36
Spittlebugs		44
Octopuses		65
Earthworms		66
Slimy salamanders		88
Sea hares		91

Banana slugs	✳✳✳❗	105
Sirens	✳✳✳	6
Snails	✳✳✳	122
Bone-eating worms	✳✳✳	97
Worm-snails	✳✳✳	18
Giant African land snails	✳✳✳	39
Polyester bees	✳✳✳	41
Lancet liver flukes	✳✳✳	62

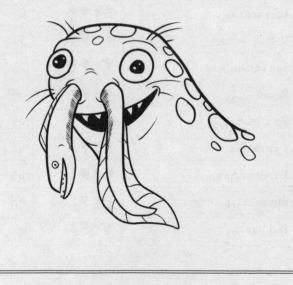

Eels	✳✳✳	69
Parrotfish	✳✳✳	70
Cereal leaf beetles	✳✳✳	74
Sea lions	✳✳✳	76
Maize	✳✳✳	81
Violet snails	✳✳✳	9
Frogs	✳✳✳	110
Bivalves	✳✳❶	19
Slime heads	✳✳❶	22
Tubelip wrasses	✳✳❶	73
Box fish	✳✳❶	118
Upside down jellyfish	✳✳❶	121
Water fleas	✳✳❶	47
Cave glow worms	✳✳❶	61
Lungfish	✳✳❶	46
Greater sac winged bats	✳✳❶	82
Sticklebacks	✳✳❶	100
Hedgehogs	✳✳	1

Slippery elm	✳ ✳	4
Biofilms	✳ ✳	5
Corals	✳ ✳	12
Opossums	✳ ✳	10
Myxozoans	✳ ✳	13
Fish	✳ ✳	96
Giraffes	✳ ✳	17
Bleeding tooth fungus	✳ ✳	20
Hyenas	✳ ✳	29
Hippopotamuses	✳ ✳	33
Horned lizards	✳ ✳	40

Swiftlets	✳✳	51
Iris borer moths	✳✳	53
Genets	✳✳	55
Viruses	✳✳	58
Bristle worms	✳✳	57
Witches' butter	✳✳	63
Sea turtles	✳✳	75
Antarctic limpet	✳✳	79

Tasmanian devils	☀ ✳	95
Cows	☀ ✳	99
Sharks	☀ ✳	87
African wild dogs	☀ ✳	106
Ghosts	☀ ✳	109
Common nighthawks	☀ ✳	113
Pitcher plants	☀ ✳	107
Leeches	☀ ✳	67
Dogs	☀ ⌡	25
Stinkpots	☀ ⌡	32
Slow lorises	☀ ⌡	34

Pygmy sperm whales	✹)	85
Pangolins	✹)	14
Birds	✹	23
Great apes	✹	28
Whales	✹	52
Penguins	✹	86
Giant pandas	✹	101
Guanacos	✹	114
Dik diks	)	78
Tardigrades		49

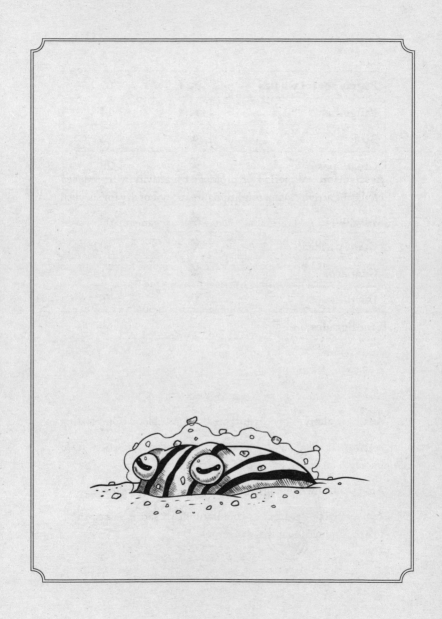

GLOSSARY

Aestivation – A period of prolonged inactivity with reduced metabolic activity when conditions are too hot or dry for survival

Amoeba – A unicellular eukaryotic organism that uses pseudopods for movement and feeding

Ampullae of Lorenzini – Organs found in some fish like sharks or lungfish that can sense electrical fields in the water

Anaphylactic shock – A severe allergic reaction that results in low blood pressure

Androgens – A group of hormones found in vertebrates that regulate male characteristics

Anosmic – The inability to sense odours

Anticoagulant – Any chemical that inhibits blood from clotting

Antiseptic – A substance that can prevent infection from microorganisms on living tissue

Aperture – An opening; in snails the opening of their shell

Aposematic – An animal's colouration that serves as a warning to other animals that it is dangerous

Astringent – Any chemical that can constrict tissues

Atromentin – A chemical naturally occurring in some fungi, which can inhibit bacterial growth and blood clotting

Bioluminscence – Light produced by a living organism

Brachial – Referring to the arm

Cephalopod – Animals, like squid or octopuses, that are in the class Cephalopoda

Corallivores – Animals that eat coral

Chromatophores – Cells that contain pigments reflect light

Ctenidia – The gills of a mollusc that serve as the respiratory organ

Desiccate – Causing something to be dry by removing moisture

Elasticity – A material's ability to keep its shape after being stretched or compressed

Ephemeral – Something that is not permanent

Epiphragm – A structure created by snails consisting of dried mucus surrounding the opening of their shell that helps to prevent moisture loss

Frothing – Churning a liquid so as to cause the production of bubbles or foam

Gastropod – Animals, such as snails and slugs, that are in the class Gastropoda

Gelatinous – A substance with a jelly-like consistency

Harem – A social organization of animals that consists of one male (sometimes two males) and multiple females

Hydrothermal vents – Water that is heated by volcanic activity and flows out of openings in the sea floor

Incisors – Teeth that are adapted to cut, found near the front of the mouth

Malodorous – Having a bad smell

Maxillae – In invertebrates, the chewing portion of the mouthparts

Metamorphose – Change of an animal's form into an adult stage

Microorganisms – Any organism that cannot be seen without the aid of a microscope

Mollusc – Any animal within the phylum Mollusca such as snails or squid that has a cavity, known as a mantle, used for breathing and excretion

Mucins – Proteins found in mucus that have the ability to form gels

Mucophagy – Feeding on mucus

Mycologist – A scientist that studies fungi

Myrmecophagy – Feeding on ants or termites

Nematocysts – A cell that can fire a projectile, which can be barbed or venomous

Neurotoxin – A substance that can inhibit or be destructive to the nervous system

Noxious – Something that is harmful or unpleasant

Obligate – In biology, a species that is restricted to a particular habitat or mode of life

Phagomimicry – A type of defence in which the secreted chemicals imitate a food source for a predator

Pinnipeds – Seals and sea-lions

Pheromones – Any chemical produced by an animal that can affect another of its species' physiology or behaviour

Pneumostome – The breathing structure of a snail or slug

Polysaccharides – A molecule consisting of many sugars bonded together

Pseudopods – A projection of the cell membrane that isn't permanent

Reticulum – The second compartment of a ruminant's (such as a cow's) stomach

Rhinorrhea – Commonly called a 'runny nose', a condition that involves the filling of a nasal cavity with mucus

Rhinotillexis – The act of removing nasal mucus with a finger

Rhinorrhea – Commonly called a 'runny nose', a condition that involves the filling of a nasal cavity with mucus

Rumen – The first compartment of a ruminant's stomach that initially receives the swallowed food or cud

Simian – A clade that contains apes and monkeys

Suprapedal – The gland of gastropods that is located above their foot and produces mucus

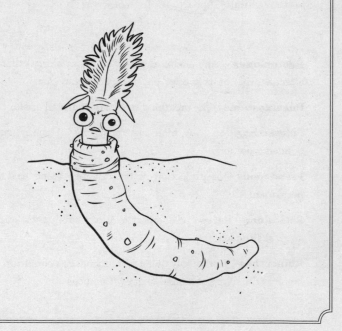

Uropygial gland – Found at the base of a bird's tail that secretes a waterproofing oil

Vestigial – part of the body that has been reduced in the course of evolution

Viscoelastic – A substance that both has a resistance to flow (i.e. viscous) and can retain its shape after stretching or compression (i.e. elastic)

Viscous – A state of matter between a solid and a liquid, and describes a substance's resistance to flow

Water column – A vertical section of water stretching between the surface and floor of a body of water

Xerocole – An animal that has adaptations to resist water loss and inhabits desert environments

Zooxanthellae – Single-celled, photosynthetic protists that live with marine invertebrates like corals and jellyfish.

ABOUT THE AUTHORS

Dani Rabaiotti has always had a passion for science and was determined to become a marine biologist from around the age of 4. Despite being waylaid a little, her passion for animals remained and in 2014 she achieved her dream of getting a place on a PhD programme between the Zoological Society of London and UCL.

Dani completed her PhD on the impact of climate change on African wild dogs in May 2019. Her research spans various fields, including zoology, quantitative ecology, spatial ecology and macroecology. Whilst her research has involved fieldwork in Kenya, in reality Dani actually spends most of her time in London at a computer, coding, which may be somewhat less glamorous, but is just as fun, and considerably less stinky. She considers it of paramount importance that everyone is aware of the fact that African wild dogs have the best ears in the animal kingdom.

Since finishing her PhD Dani has been working as a freelance scientist and science communicator. At some point she hoped to follow in Nick's footsteps and get a job as a post-doctoral research assistant. Her coolest wildlife sighting was seeing dolphins riding the bow-wave of a blue whale, in Monterey, California. You can follow Dani on Twitter and Instagram at @DaniRabaiotti

Nick Caruso grew up in St. Charles, Missouri where his love for the natural world and fascination with animals began. He spent much of his childhood catching all manner of reptiles and amphibians while playing in forested areas, creeks, and streams with his brother. He has since turned that passion into a career by studying the ecology of those same species that captivated him as a child. After receiving a BA degree in biology from Saint Louis University, he completed an MSc in biology from The University of Maryland College Park and a Ph.D. in biological sciences at The University of Alabama.

Nick is currently a postdoctoral associate in the Department of Fish and Wildlife Conservation at Virginia Tech. His research focuses on quantitative population biology and conservation of amphibians and reptiles. He currently works with both Appalachian salamanders and herpetofauna in the Floridan panhandle. When he is not researching amphibians and reptiles or writing best-selling books about farts, he's mountain hiking with his wife, weight lifting, playing racquetball or soccer, or cuddling his cats.

You can follow Nick on Twitter at @PlethodoNick

ABOUT THE ILLUSTRATOR

Ethan Kocak is an artist and illustrator best known for the online comic series *Black Mudpuppy* and various science-related art projects including the *New York Times* Bestseller *Does It Fart?* and TV presenter and biologist Ben Garrod's dinosaur series *So You Think You Know About Dinosaurs?* He lives in Syracuse, New York with his wife, son and collection of rare salamanders.

CONTRIBUTORS

Thanks to our various slime experts, who contributed to this book. They can be found on twitter at:

Elly Knight @ellycknight
Jack Ashby @JackDAshby
MaLisa Spring @EntoSpring
Catie Alves @calves06
Maxime Dahirel @mdahirel
Elizabeth Ostrowski @elizostrow
Jacinta Kong @jacintakong
Christopher R Wright @WilyMongoose
Crystal @maneatingplants
Rachel Hale @_glitterworm
Sarah McAnulty @SarahMackAttack
Tori Roeder @Tori_Roeder
Megan McCuller @mccullermi
Alexander Robillard @AJRobillard
Katie O'Donnell @katie_m_o
Christine Cooper @CECooperEcophys
Alexander S H Dean @Fungiguy2
Fraser Januchowski-Hartley @Nasolituratus
Débora @debnewlands

Dennis @TheBushFundi
MarAlliance @MarAlliance
John Tulloch @JT_EpiVet
Nicola G Kriefall @BiologyForLyfe
Bailey Steinworth @baileys
Jessica Light @je_light
Justine Hudson @justinehud
Vanessa Pirotta @VanessaPirotta
Audrey Dussutour @Docteur_Drey
Robert Insall @robinsall
Damaris Brisco @fungal_love
Abid Haque @abidhaque
Hailey Lynch @HaileyLynch99
Subash K Ray, The Swarm Lab, NJIT

First published in Great Britain in 2019 by Quercus.
This paperback edition published in 2020 by

Quercus Editions Ltd
Carmelite House
50 Victoria Embankment
London EC4Y 0DZ

An Hachette UK company

Copyright © 2019 Dani Rabaiotti & Nick Caruso
Illustrations © Ethan Kocak

A CIP catalogue record for this book is available
from the British Library

PB ISBN 978 1 52940 340 4
Ebook ISBN 978 1 52940 338 1

10 9 8 7 6 5 4 3 2

Design by Rich Carr, Carr Design Studio

Printed and bound in the UK by Clays Ltd, Elcograf S.p.A.